D0629971

THE NEXT APOCALYPSE

CHRIS BEGLEY

THE NEXT APOCALYPSE

THE ART AND SCIENCE OF SURVIVAL

BASIC BOOKS
New York

Cover design by Ann Kirchner
Cover image © Seita / Shutterstock.com; © Nosyrevy / Shutterstock.com; © brushpiquetr / Shutterstock.com

Basic Books
Hachette Book Group
1290 Avenue of the Americas, New York, NY 10104
www.basicbooks.com

Printed in the United States of America

First Edition: November 2021

Published by Basic Books, an imprint of Perseus Books, LLC, a subsidiary of Hachette Book Group, Inc. The Basic Books name and logo is a trademark of the Hachette Book Group.

Print book interior design by Amy Quinn

Library of Congress Cataloging-in-Publication Data
Names: Begley, Chris, author.
Title: The next apocalypse : the art and science of survival / Chris Begley.
Description: First edition. | New York : Basic Books, Hachette Book Group, 2021. | Includes bibliographical references and index.
Identifiers: LCCN 2021016180 | ISBN 9781541675285 (hardcover) | ISBN 9781541675278 (ebook)
Subjects: LCSH: Survivalism. | Disasters—Social aspects. | Emergency management—Social aspects.
Classification: LCC GF86 .B435 2021 | DDC 613.6/9—dc23
LC record available at https://lccn.loc.gov/2021016180

ISBNs: 978-1-5416-7528-5 (hardcover), 978-1-5416-7527-8 (ebook)

LSC-C

Printing 1, 2021

For Soreyda, Bella, William, and Aaron

CONTENTS

PROLOGUE

Nothing lasts forever. The laws of physics demand we descend toward disorder, and we comply. Every change sets in motion the next until the transformation permeates our lives and communities. These changes can be profound and dramatic—sometimes apocalyptic. Many of us think about these apocalyptic changes, a lot. If it is not upon us, we convince ourselves that the next apocalypse is just around the corner. I suspect every generation feels the end is near. Sometimes they are correct.

The effects of profound changes endure, but we also remember the world before it transformed. We create a mythical past, a romanticized time when everything was in its proper order. I am an archaeologist, and I study how societies form and transform. In the rainforest in Honduras, I saw how profoundly these changes affect people, even centuries later.

"Do you know the story of the Lost City?" Don Cipriano asked me. "It's right up there." He gestured with his hand.

Cipriano and I were sitting and looking into the campfire, waiting for water to boil for coffee. We were in the Mosquito Coast of Honduras, deep in the rainforest, where I had conducted archaeological research for years. We had been walking for over a week, documenting the archaeological sites that he would show me. I had heard the stories of the Lost City, of course. Everybody recounted the Lost City story, and many people claimed to have found it.

"Should we go see it?" I asked, skeptical but interested in the opportunity to look at another ancient site.

"No," Cipriano answered. "We can't. It's where all the gods fled when the Spanish came. Our gods, and from all the original people." Cipriano is Pech, an indigenous group descended from the people who built the archaeological sites a millennium ago. I worked with the Pech because they knew the rainforest better than anybody else did. I had been living in his village for a year or so. Cipriano continued, "The gods are lonely, and you must talk to each one if you visit the city or they get angry and never let you leave. That means we would have to know all seven languages. We don't." Nobody knew all seven indigenous languages.

"Have you been there?" I asked. Cipriano looked at me, barely concealing his frustration and impatience.

"No. Did you not understand the part about the languages? If I had ever been, I would still be there. But I know where it is." He paused, and his shoulders sank as the irritation, even his energy it seemed, disappeared. His voice trailed off.

"Maybe we can go someday," he said. "I don't know . . ."

The sudden resignation surprised me. He wanted to go, it was clear, but he knew he could not. Going there would be like traveling back to an idealized time. He could never go back.

We never spoke of it again, and I never visited that place, but I understand why it was so important to him. This lost-city legend, the legend of the White City, is centuries old. Spanish versions have been conflated with the Pech stories. Today, the stories describe a large, once-wealthy city lost in the jungle. In the original Pech stories, it is not a city. They use the term *Kao Kamasa*, or White House. Unfamiliar with the Pech language, the name was mistranslated by outsiders. No hint of scale or wealth appears in the original story, because it is not important. The important part is that this place is home to the gods and located in the traditional homeland of the Pech.

There are many impressive archaeological sites in the area, but this one represents his people's ancestral home. This is where they were before their traditional world collapsed and they were scattered by the interloping Europeans. This represents society before it fell apart. From Cipriano's reluctance to visit the place, and his desire to talk about it, I realized that what was important was not the place but what was lost: the autonomy and the dignity that could never be found in the rainforest. It was not about the place: it was about a mythical, half-remembered, half-imagined, pre-collapse past.

For Cipriano, this distant event is not just about the past. His image of a mythical past formed the framework around which he constructs his present. For him, the present is a distorted and grotesque version of what might have been—what would have been—had it not been hijacked by interlopers. If this image of a distant, pre-collapse past has that power, what about our visions of a post-collapse future?

How we think about the future is important; it shapes that future. As Cipriano's vision of the past affects his view of the present, how we think about the future leads us to act in certain ways now, and in turn, those actions affect the future. When I look around now, I see us imagining an apocalyptic future. We write about it, make films about it, and prepare for it. We think about things falling apart. We create elaborate apocalyptic fantasies, which are both painful and pleasurable to indulge. Those fantasies inform our preparation for disaster and set the parameters of possible futures.

Why do we do this? Maybe we are worried about the future, understandably. We face enormous challenges, with more on the horizon. It seems to be more than that, however. We seem to enjoy the fantasizing. Our fantasies reflect our history, desires, and fears. These fantasies shape the stories we create. Those stories influence our future.

Like most archaeologists, I study societies that are constantly transforming, including some catastrophic transformations we label "collapses." Archaeologists study how and why these changes occur, and how people react to rapid or profound change. This topic is familiar to most archaeologists and, for some, central to their research. I look at what actually happens when things fall apart, and I see where our imagination gets it wrong. Our fantasies are not consistent with how any of this has ever played out, in any situation. Apocalyptic narratives in books and movies look nothing like real catastrophes. We have an entertaining and convenient but inaccurate view of how these things happen. We are preparing for the wrong disaster.

INTRODUCTION

I am an archaeologist, but I also teach wilderness survival courses. This interest grew out of my archaeological research in remote areas of Central America. I learned many of the skills I teach during the years I lived and worked with people from the Pech indigenous group in Honduras. Skills like making fires were daily tasks in the village. In fact, many of the survival skills I teach were everyday activities, especially when we were camping in the rainforest during our many trips documenting archaeological sites. In my survival courses, we discuss disasters, emergencies, and the unexpected. People want to know how to prepare, and how to survive. Although I inhabit the world of academics, the world of preppers and survivalists is only a step away.

In the last few years, there has been an increased interest in my wilderness survival courses. Originally, I'd envisioned these courses as preparation for unexpected, short stays in the outdoors, such as getting lost while hiking. What I find, however, is that people want to learn these skills in order to be prepared for a large-scale disaster. They want to get ready for pandemics, economic collapse, or the rise of authoritarianism. The disasters they imagine reflect increasing fears of an unstable world. Until recently, climate change drove these fears. Now, the fears include pandemics or political unrest. My students asked me about my own escape plan. What am I going to do? What should they do? How do we survive the next apocalypse?

This book stems from those experiences. I look at past events, to see what radical change actually looks like. I examine our current fantasies about the apocalypse, reflected in our art and our preparation for disaster, revealing our hopes and fears. I explore how we are shaping the future through our framing of possible scenarios, and how our reaction ensures the replication of our current situation, of the status quo. I look to the future, and likely apocalyptic scenarios, discussing the skills and actions that help us through these events. We always get the future wrong. Sometimes it comes out of left field. Nevertheless, by imaging possible futures and by understanding how we shape and limit our own vision of the future, we can see it further out, buying us time to react.

I did not write this book as a response to the Covid-19 pandemic, but I wrote it during that pandemic, and the contents reflect that. I began months before anyone had heard of the virus, when climate change seemed to be the major threat to the world as we know it. Seeing narratives of dystopian futures everywhere, from film to books to video games, and being familiar with how and why societies changed in the past, I recognized the significant differences between reality and our fantasies. We create and consume apocalyptic and dystopian narratives. Every other young adult novel is a dystopian fantasy. Zombies are everywhere. We love to think about the next apocalypse. I wondered if our imaginary apocalypses resemble anything that has ever actually happened before, and if our fantasies affect how we think about the future, how we prepare for disasters, or even how we act in the present.

I published an op-ed in my local newspaper on that theme, suggesting that our fantasies reflected our celebration of individualism and self-reliance.[1] They reveal our desire to return to a mythical past that never existed, but this imagined future does not resemble any type of actual apocalyptic aftermath. That essay resonated with many people, and I expanded it into this book. Then the pandemic hit, and apocalyptic conversations shifted.

I want to examine our ideas, compare them with data we have, and move on from there. We do not want to go forward planning for an imaginary problem while ignoring more evidence-based models of challenges to come.

In Part I, I examine the accuracy or plausibility of our ideas about an apocalypse. In order to assess this, I look into the past. As an archaeologist, I deal with some societies that have disappeared, some that have radically changed, and some that seem to have great continuity over time. Do we see anything like our apocalyptic fantasies in the past? Do societies really collapse, as we like to imagine? If so, why? What does it look like? What can we learn from that? I look at archaeological and historical evidence of a few events that represent the types of catastrophes we imagine, and compare those to our apocalyptic visions.

In Part II, I look at our contemporary focus on the apocalypse. Why are there hundreds of books, movies, and websites that tell apocalyptic stories or talk about how to deal with these events? When did this obsession begin? What ideas about the apocalypse do these works reflect? What are the reasons that the apocalypse remains a popular topic, and how does this change over time? What can we learn about ourselves by looking at all this? How do people prepare, and why? What do they imagine will happen? What do they think about the future?

In Part III, I ask what this means for the future. Could our society collapse? If it did, what would likely cause that? What would it look like? What could we do to survive? I explore the most likely ways in which some sort of apocalyptic tragedy could befall us, and then I consider some unlikely scenarios. Finally, I ask what we can do to make it through those events.

Ultimately, this is not a doomsday book. I am not interested in fearmongering, lamenting our contemporary reality, engaging the fantasy of some mythical past, or advocating the importance of skills from a simpler time. I want to explore what actually happens in a catastrophe, how we make it through, and why we think about all of this in the ways we do.

PART I

The Past

Archaeology of the Apocalypse

In 1994, I arranged with the Field Museum in Chicago to bring back a collection of modern ceramics from the indigenous village in eastern Honduras in which I lived while conducting archaeological fieldwork. The village, Nuevo Subirana, is a Pech village of about 500 people, although it seems much smaller, spread out along the valley floor. I was identifying and documenting archaeological sites throughout a series of river valleys. There were many villages in which I might have lived during the project, but I decided to ask the Pech if I could rent a place in Nuevo Subirana and if anybody would be interested in working with me on this project. The community granted permission, and a half dozen people worked in the field with me. Many people in the community used ceramic pots to cook in—they were preferred for certain things, like beans, much in the way we are fans of old-fashioned cast-iron skillets here in Kentucky.

In my archaeological work, we find pottery more than any other type of artifact. Almost all of these are broken sherds, but I can identify the size and shape of the vessel, understand the particulars of the shape and design, and see how the decoration of the pottery changed or endured over time. Archaeologists typically care more about the information we can get from an artifact than the state of the object, so broken pots are valuable to us. I found a suite of design elements recurring on the pottery across a wide region of eastern Honduras. Almost all were carved or pressed into the surface and consisted of recognizable formations of dots or lines.

It was common to see broken pieces of modern pots outside of houses. Some pots lasted decades, but most did not; people routinely broke them accidently. I asked my neighbors about these ceramic pots. What did they cook in them? Were they preferred for certain things? How long do they last? How did they typically break? How do you dispose of the broken pots (a strange question for anybody but an archaeologist, perhaps)? I learned that a woman named Gloria made all the ceramic pots in the community. She lived at the other end of the village, about a kilometer away from my house, and I rarely saw her. I visited her and talked about pottery. I asked if I could buy around fifty pots for the Field Museum, and she agreed to make them over the next month. I arranged to come back and talk with her about her process for the report that would accompany the collection to the museum.

A couple of weeks later, I returned to her house. She showed me the pots she had made. I noticed the decoration around the rim that she had added to several of them. They were nearly identical to the designs on the thousand-year-old sherds we found during our archaeological work. This was an exciting and potentially significant development. Did she know something about these designs? Did she know what they meant? This potential evidence of some continuity from the distant past to the contemporary Pech could be significant. A direct line from the pottery-making of her ancestors to Gloria's own practice would be remarkable. It would also be invaluable to our understanding of the past. Sometimes archaeologists can interpret the iconography from the past, but in this region, I do not know what the markings mean or represent. I asked her about the patterns.

"Well, you know the broken pieces you find in all the fields around here? I figure those were made by our ancestors, so I put them on my pots too," she said. She laughed and shook her head when I asked if she knew what any of the patterns meant. I was disappointed, but I also realized that she *had* explained what the patterns meant—at least what they mean now, to her. They signaled a desire to maintain a connection with a distant past, one that had been radically shattered by European colonization and the subsequent racism, poverty, and marginalization that came to shape life as an indigenous person. She revealed to me my limited vision of "meaning" and suggested that she engaged with the past in a different way, one that only sometimes intersects with the way an archaeologist might see it. Archaeology often reveals as much about the archaeologists as it does about the past.

In that part of Honduras, the folks who lived a thousand years ago and the Pech are connected. At the very least, there is some connection because they occupy the same geographical space. They practice similar types of agriculture, and they speak their own unique language. In other ways, though, the continuity is less clear. There are few stories about what happened in the distant past, like Cipriano's story of the White City, and no written records before the sixteenth century. In some cases, archaeology is the only way we can access the past.

In thinking about the next apocalypse, I focus on the future. Looking at the past helps us see that future. When we look back into the past, we have a much wider lens, and a far greater sample of events to observe than that which exists in the present. When we are looking to the past for guidance for the future, there are some caveats. Obviously, some important things have changed. The global population is much greater, as is population density in most places. Our daily activities are radically different from folks in preindustrial times. Nevertheless, similarities exist. Now, as in the past, we have very complex, interwoven systems, and we may not be able to anticipate the ripple effect of change in any given part. Everything exists in a unique context, and that particular history is never reproduced. Looking to the past requires careful consideration of the ways in which the situations are fundamentally different, as well as the similarities they share. With that caveat, looking to the past sharpens our vision of the future.

Our understanding of the past is imperfect. Archaeology, like all science, is a process. In the face of new data, conclusions might change. Every new study adds to the previous, and sometimes generates data that refutes previously held ideas. In the next chapters, I mention some conclusions proposed by archaeologists that we no longer think to be true. This is how science works, and it should not suggest something wrong with the scientists or with the process. We make hypotheses, look at data, and come to conclusions based on the data we have, which we think is sufficient to posit whatever we are concluding. Eventually, we collect new data, or we look at the data from a new perspective, and we change our ideas about which hypotheses are supported. We do not all agree always, of course. Scholars argue all the time, about nearly everything, but that drives the process; it is not a sign of dysfunction. New generations of archaeologists will improve upon my work in the Mosquitia of Honduras, for instance. They will build on my research with more and better data, cleverer interpretations, or fresh perspectives, and will come to a better picture of what happened in the past. We are always moving to new understandings, and it is not an indictment of previous researchers or of the discipline. Scientists build a stairway. We add steps to allow the next in line to climb higher, and see further. The particular step we build is not the accomplishment; the stairway is.

Many factors make it difficult to access and interpret the past. Sometimes it is field conditions, or fragmentary data. Sometimes the issue lies with the archaeologist, or with the discipline of archaeology itself. A major challenge for archaeologists is our colonial legacy. Archaeology, and our parent field, anthropology, developed during and in concert with the colonizing of the world by European powers. In many hidden ways, that situation colors how we behave and how we think about the world. Archaeology has a long history of colonizers studying the past of colonized people. This continues to this day. I am from the United States, and I study the past in formerly colonized parts of the world. This is not unusual. What is much less common, however, is for archaeologists from a formerly colonized part of the world to be conducting archaeological research in the metropole, the home of the colonizers. It happens, but rarely. These patterns are only one marker of the colonial legacy in archaeology.

The colonial legacy that undergirds archaeology continues to shape its practice. It affects who participates in archaeology, what questions we ask, what traces of the past we value and interpret, and what we think constitutes data or proof. It influences what we produce as archaeologists, and how we present our findings. Who participates in archaeology might be the first thing we need to change, so that a more diverse group can decide how to address the other problematic and limiting vestiges of this colonial history.

The privileging of one group over others means that a limited slice of humanity writes the past for the rest of us. My vision of the past resides in the present, obviously, and reflects my lived experience. Like all archaeologists, contemporary power dynamics and the systems in which we live shape my understanding of the past. This includes the unjust and inequitable parts of that system. I must consider how those realities could influence interpretations of the past. In my case, for instance, I was born into the late-twentieth-century United States of America. My experience is that of a white male, living in a patriarchy built on a white supremacy and the exploitation of Black and Brown bodies. As much as I might abhor elements of my group's history, I cannot escape it, and my understanding of the past will never be completely divorced from a particular perspective.

I try to eliminate bias, to minimize the parallax from my position, but I will always fail. I could cite many examples of things that shape my perspective and bias my outlook on the world. For instance, I default to a binary gender system if I do not watch myself, even though I know that people across space and time have had a much more complex way of thinking about gender. As another example, I notice that the motivations I attribute to people reflects my upbringing in a capitalist system that celebrates individual achievement. I have internalized ideas about how people behave, and what drives their behavior. I have to make a conscious effort to imagine a group who would not celebrate individual achievement in the same way. I know that societies exist in which the larger group is the focus, where celebrations focusing on the individual accomplishments would be construed as embarrassing, selfish, or vain. This variation in worldviews exists right now, within the limited parameters of our contemporary world. Imagine if we extend that back 300,000 years to the beginning of modern humanity. An interpretation of the past benefits from a variety of perspectives, in part to eliminate or minimize biases and blind spots. The history of archaeology has limited the number of perspectives participating in the interpretation of the past. More and varied vantage points strengthen our understanding, and the content of this book benefits from the range of voices included here. Archaeology in the future will benefit from an even greater range.

CHAPTER 1

WHEN THINGS FALL APART

There are many examples of events commonly labeled "collapses" in the ancient world, from Easter Island to Babylon. Eventually, everything transforms or disappears. I focus here on three examples of profound change in the past in order to explore what these events looked like, what caused them, and what it meant for the people going through these transformations. I look at well-known "collapses" in areas I have worked in; the Classic Maya civilization in Central America and Mexico, the Western Roman Empire around the Mediterranean, and the many Native American societies in eastern North America after the arrival of European colonizers.

I've worked in many different parts of the world as an archaeologist and have had the opportunity to explore the realities of past apocalypses. I look at what happens when societies collapse, and ask if that concept, *collapse*, is accurate. Archaeologists often find the term "collapse" problematic. These issues, which had been voiced in various ways over the decades, found a focal point with the publication of Jared Diamond's *Collapse: How Societies Choose to Fail or Survive*.[1] While acknowledging the value in Diamond's work, especially his ability to take research and present it to the public in an accessible way, many archaeologists felt that his analysis of particular case studies was not compelling, and that the insistence on finding collapse in each of these cases was problematic. Several colleagues have worked on projects examining the concept of collapse, and some have compared past events with the present and future, as I do here.[2] I assume that somebody else has already had every clever thought I have, and I have not been wrong yet. In this case, many archaeologists have written about the idea of collapse, that being an important and problematic topic for us.[3] I reached out to several of those archaeologists, and I am grateful for their input.

Many of the events we label collapses may not, in fact, be collapses. Certain elements of a society change, a political system changes, or people spread across the landscape in new ways, but in very few cases did something collapse so thoroughly that it is unrecognizable. While certain elements of the Classic Maya way of life largely disappeared, millions of Maya live in Guatemala, Honduras, Belize, and Mexico today, and their language, religion, and certain everyday practices exhibit a connection with the distant past. Our labels, like the term "collapse," can lead us to picture the past in an inaccurate way. We have to be very careful in our interpretation of archaeological data.

To an archaeologist, change can look very dramatic and sudden. It can appear that something was here one day and gone the next, when in reality we might be looking at changes that took place over a century. The type of fine-grained temporal resolution that we need in order to understand how a particular society changed is often not available to archaeologists. We have techniques like Carbon-14 tests to determine the age of something, but there is always a margin of error based on factors like sample size. In practice, we mainly use the style of the artifacts and architecture that we find to tell us the date. Styles change over time, and we recognize the age of an item much as we can tell the age of a car from its shape and design, or the age of a photograph by the haircuts and clothing worn by the people in it. In my own research, I often cannot narrow things down beyond several decades. This is common, especially in places that have seen little archaeological research. Even in places where an enormous amount of research has been conducted, the resolution is still relatively coarse. That can affect how we interpret what we see.

Because of this phenomenon, archaeology tends too easily to identify changes in style or design as evidence of disruption or discontinuity in a culture itself. The weight we give to the presence or absence of something affects our interpretations. If we use certain markers to identify a specific group, like the monuments they create in a certain style, or how they record dates, we see it as significant when those things change because we use those things to define that group. When those defining, diagnostic behaviors disappear, it follows, in our mind's eye, that the group must also have disappeared. In some ways, I am looking at continuity versus discontinuity in a society. An apocalyptic event would sever continuity. As archaeologists, we see changes in material culture as evidence of discontinuity when, in fact, this type of significant severing of past from present might not have occurred.

In my experience, I can think of technological changes—even significant ones—that no one today would see as creating a fundamental break with the past, even if an archaeologist from the future might make such an interpretation. Many changes have little effect on the larger society. Technological changes, such as the switch from manual transmissions to automatic, had very little impact on the larger culture outside of our driving experience, although they were sometimes touted as revolutionary. Even rapid and widely felt changes, like the switch from vinyl to CDs and then streaming music, do not represent the kind of discontinuity I might imagine if I were viewing them in an archaeological context. While these changes certainly affected the music industry, and may have changed our relationship to music, they do not represent a fundamental societal change.

In the past, the disappearance of a certain type of monument, or a change from construction of one sort to another, may not be as significant as it seems to an archaeologist interpreting that change. This tendency to create an apocalyptic narrative out of the past is a particular bias of archaeology, where we perceive discontinuity too readily. The ways in which societies endure and continue can be obscured by this tendency to lean toward discontinuity. This is understandable; we notice new things. We notice things that are out of place, or recently changed: the discontinuity. We do not foreground the unchanged things: the continuity. This may not be limited to archaeology, however. The visions of discontinuity that come so easily color our interpretations of the past as well as our visions of the future.

We must also consider the names we use for these changes. Not all change is bad, of course, and change is inevitable. Using the term "collapse" to describe a particular set of changes suggests failure; something was standing, then some part of it failed and it collapsed, like a tower. It follows that a collapsed society is a failed society. However, a society that has changed might not have failed, necessarily; it might have changed for the better. Some political systems actively work to disadvantage or disenfranchise part of the population. For those people, a collapse would not be negative. Many archaeologists have identified collapses that empowered some, or even most, of the population. From the collapse at the end of the Bronze Age in the Mediterranean to the civilizations of the Bolivian altiplano, archaeologists see the collapse of an existing system as potentially revolutionary, as a way for another system to arise.[4] Therefore, we must be careful in equating the type of "collapses" that we identify in the ancient world with a truly apocalyptic event, which would be overwhelmingly negative. We see that our vocabulary matters, and the concepts we use to think about these things can change the way we understand the past.

Archaeologists need to think about why we call some past transformations collapses, implicating the past group in a kind of failure, while similarly significant transformations for other groups are not labeled in that way. During a recent conversation about the terms we use, archaeologist Patricia McAnany commented, "We don't go to Stonehenge and say, 'Wow, these people must have really screwed up their environment. Why did they leave? Why did they stop using Stonehenge?' No, we don't say that. We just think, Well, there was something going on here that's probably kind of astronomically based, and then at a certain point it was just not important anymore, except maybe as a pilgrimage locale." Not every group gets the same treatment. Other groups collapse; when it's us, we simply change.

"Apocalypse" is not a word archaeologists use to describe past catastrophes, typically. We often use the word "collapse," but we do not agree on the definition or usefulness of that concept. Some scholars reject the term, arguing that "collapse" signals something more drastic, complete, and sudden than the realities of the situations merit. Not all archaeologists reject the concept of collapse, however. Some think that other terms like "transformation" may be misleading and not reflective of the trauma and horror of the lived experience of people going through it. Sometimes the events of the past are so profound and rapid that the term "collapse" is appropriate; when I use the term, I mean it to signal the speed and significance of a transformation.

In order to explore this issue further, given the disparate opinions evident in the literature, I contacted several archaeologists who worked directly with these issues, to ask specifically about their ideas of a collapse, and how this might inform us about the future. I use this quasi-journalistic approach here because it allows these scholars to expand on published works, or to talk about these concepts from new angles. One of the first archaeologists I thought of was Patricia McAnany, whom I just mentioned. She teaches at the University of North Carolina at Chapel Hill, works in the Maya area, and writes extensively about the concept of collapse in archaeology.

We talked about some of the difficulties that archaeologists have in coordinating their data with behavior in the past. It can be difficult to fit our carefully crafted theories about the past with the incomplete and coarse-grained data we recover. We make up for this lack of resolution by the breadth of our vision. We can look at very long-term trends. Archaeologists recognize the power in our ability to look diachronically, over time. However, challenges remain. We do not always know the cause of long-term trends, and our interpretations are influenced by the challenges of our contemporary society.[5]

McAnany understands that our ideas about the past tend to reflect the present, and that we transfer our current concerns to our interpretations of the past. In the 1960s, for instance, she says, "The Vietnam War was raging, and so Mayanists were starting to talk about war as a cause of collapse. And then when religious fundamentalism became an issue [in the 1980s], then religion started to be talked about as a possible cause of what happened to those southern cities in the ninth century, why so many of them were abandoned, why the population decided to go elsewhere."

Archaeologists identify various causes for collapses or transformations. Sometimes, the actions of a particular group seems to have brought about the catastrophe. In many other cases, however, we see something outside the group as causal. If we are not careful, we envision societal collapse in the past as something that happens *to* a group, as if that external force determines the outcome of the crisis and the people themselves are passive or ineffectual in their response. Many archaeologists recognize, however, that the societal *response* to an external force or crisis can shape the trajectory of a society more than the external force itself. One example of focusing on the external force, rather than the response to it, is the use of environmental factors to explain past collapses. Archaeologist Guy Middleton, from the Czech Institute of Egyptology at Charles University in Prague, writes about this extensively, and I turned to him when I began exploring these issues. He argues that the dominant paradigm for explaining collapse is environmental.[6] Not only does this reflect our current concern with environmental policy and sustainability, this focus on external forces also "obscures recognition of the dynamic role of social processes." In other words, we forget that even widespread phenomena like environmental degradation or climate change generate complex human responses, and that these responses can shape the outcome of the crisis, perhaps more than the proximate cause, the original crisis, itself. As we shall see, the particulars of how a collapse plays out are not fully determined by the nature of the proximate cause. Rather, that cause is mediated by complex systems and how they respond as that cause reverberates through stages of transformation and reconstitution.

The length of time that a society endures, and the time involved in its decline, are often lost in our discussions. McAnany recounted a story that illustrates this: "A few years ago I visited a town in Germany because that's where Charlemagne was located. I thought afterwards that no one would think about going up to contemporary residents and asking them, 'Do you miss Charlemagne? Weren't those the good old days? Too bad Charlemagne was responsible for environmental destruction or too much warfare and the whole system collapsed.' You don't think of it in that way. You think of it that it was a very long time ago and governmental systems were construed quite differently then." Such questions *are* asked of some groups, like the Maya, related to the colonial legacy I discussed earlier.

Below, I introduce three case studies I use to explore dramatic, perhaps even apocalyptic, change in the past. These examples from Central America and Mexico, the Mediterranean, and eastern North America provide a contrast in time and space, but also in the interpretations of the events that led to the decline or collapse of each group.

THE CLASSIC MAYA COLLAPSE

It was all gone. The house, even the land. It was ocean now. I stood there, in a small village on the coast of Honduras, looking at the place where my wife's childhood home once stood. I recognized nothing, oriented only by the half of the school building that survived the flood. How was I going to tell her that the flooded river had jumped its bank and changed course, charging straight through the barrier island where the village stood, cutting a new channel right through the property? She grew up in the house that was no longer there, that her grandfather was going to pass on to her. She talked about moving back, maybe retiring there in the village where she lived as a child. But it was gone. Not just the house but all of it. It was part of the Atlantic Ocean now.

Sad as that was, it was nothing compared to the unfolding tragedy in the village. Two powerful tropical storms had unleashed record rainfall only a month apart. All the crops were gone. One man was gone, too, drowned in the raging river that had torn through the village. Two one-hundred-year storms in two months, after a once-in-a-lifetime hurricane a couple of years earlier. The weather was changing, the climate was changing, and this is what it looked like. People displaced, leaving because of the destruction of houses and crops. The village would lose a third of its population. This was the latest iteration of natural effects that had been happening for millennia. Drought, deforestation, sea-level rise at the end of the last ice age—all of these changed where and how people lived.

This all happened in part of Honduras near the scene of the best-known collapse in the Americas—that of the Classic Maya in the ninth century CE. About the time Islam spread to West Africa, when *Beowulf* was being composed and Charlemagne was battling the Saxons, people in parts of Mexico and Central America were abandoning their cities. Sometimes, we find traces of how it happened, as in the case of communities in the rainforest in Guatemala, where people stripped off the façades of their temples and constructed rubble walls as a defense against attackers.[7] Those communities had been abandoned by most, but a small group fortified the area and made a stand. Drought, environmental degradation, violence, and the collapse of political and economic systems drove people away from the once-great cities. Through all that, communities persevered. Today, more than 7 million Maya live in the region, suggesting that a focus on apocalypse can obscure substantial societal continuity, and with that a degree of interdependence that belies our lone-wolf visions of survival.

The term "Maya" encompasses over thirty groups of people. They speak different languages and live in different places, although they all have some connection. That connection, however, can be relatively distant, and this is not a homogenous group. In the face of overwhelming adversity, a unifying category arose under which they have coalesced. "Maya" is a modern term, not used by ancient groups. This modern collective term is similar to the way that the concept "Native American" became a meaningful category in the nineteenth century, whereas before, a unifying category did not exist.

The Maya area is large and varied, and includes parts of Mexico, Guatemala, Belize, Honduras, and El Salvador. Much of this area is rainforest, which I believe influences how we think about the collapse. Part of the mystique surrounding the Maya and their history relates to the fact that some of the larger Mayan cities are located in areas that are now rainforest. One of the synonyms used for rainforest is "jungle." That word evokes many images, many of them negative. "Jungle" is a term that comes from Hindi, derived from *jangal* and referring to an area that is no longer cultivated or maintained, having become overgrown and wild. The term "jungle" was coined and popularized during the colonial era, and carries connotations as some sort of dark, menacing, and foreboding place. In a colonial context, this also carried the subtext that these overgrown areas, feral and not "properly" maintained, evading the "civilizing" effects of the colonizers, must be primitive, bad, and dangerous.

In references to my own work in the rainforest, I routinely hear the forest referred to as "impenetrable jungle," and cast in a negative way. By portraying an area in those terms, it demeans the area and its residents, who must be flawed in some way to live in such a hostile, dark place. It also sets up the jungle as a challenge to overcome; it allows heroic action within its confines by interloping explorers who are proxies of a colonial past. You can portray yourself as heroic for having survived in that environment. In a lot of the sensational reporting about archaeological sites found in the rainforest, including some recent ones in Honduras, there was this clear effort to cast the "discoverers" of these sites (already known to locals, of course) as heroes. The "lost city" trope is strongly associated with the rainforest and with the Maya homeland.

Explorers of the nineteenth century, as well as some archaeologists, writers, and filmmakers even now, represent the geography of the region as something to be vanquished. Never mind that the jungle is not a particularly hostile environment and is home to millions of families. Thousands of little kids, literally, are playing out in the rainforest right now. That inconvenient truth is not part of the representation of the region as exotic and dangerous. Since the Maya sites are so unique and impressive, and since they occupy a place often misrepresented as perilous and heroic, these archaeological sites receive a lot of attention.

The area in which I worked in Honduras, the Mosquitia, always got more attention than nearby areas with equally impressive archaeological sites because it was "jungle," sparsely populated, and thus a perfect setting for living out colonial-era fantasies, such as finding a lost city.

Despite its reputation, it is actually easy to survive in the rainforest. If I had to pick an ecozone in which to be lost, I would pick the rainforest without any hesitation. Part of that probably has to do with my familiarity with it. Another part of that preference, however, has to do with the presence of plentiful water and easily identifiable, edible plants in an area with little danger of cold or exposure. There are challenges, to be sure, including the many poisonous snakes and tropical diseases like malaria and dengue fever. Even with those, the rainforest ends up being one of the easiest places to survive. Its dark and foreboding image is not a reality; rather, it is what we see through the filter of a colonial past.

When thinking about the Maya area in the past, we should keep in mind that the archaeological sites that are now isolated deep in the rainforest would once have been large towns or cities surrounded by agricultural fields. It would all have seemed very different. Where I worked in the Mosquito Coast of eastern Honduras, some sites are now a four or five days' walk through the rainforest and are very hard to find in the midst of the vegetation. We know that in their heyday, these areas would have been cleared of forest cover for the village and surrounding agricultural fields. Rather than isolated, these sites would have been connected to others by cobble paths. We still find sections of these paths in the rainforest in eastern Honduras that would have linked sites that now seem incredibly isolated.[8]

The "collapse" that I discuss here happened between 750 and 900 CE, at the end of what we call the Classic Period that dates from 250 to 900 CE. At the end of that period, Maya cities in the southern lowlands declined in population, the production of dated monuments stopped, and larger sites were abandoned. The cause or causes of all of this are the subject of much contention. Some places were affected to a greater degree than others, and some were affected earlier. The Maya never disappeared, of course. They moved. Certain sites to the north gained strength during the "collapse," and we see people and power move north from the lowland rainforests in and around Guatemala to the Yucatán Peninsula of Mexico.[9] People also moved from urban to rural areas, in patterns that we do not yet understand.

The term "Classic Period" comes from the Europeans who arrived in the region some six hundred years after it ended. They noticed and commented upon the remnants of the Classic Period settlements, including their large buildings and impressive artwork. Certain features of the impressive archaeological sites, including monuments with glyphic texts, resonated with existing beliefs that cultures would reach a zenith in terms of creativity, power, and everything else, and then decline in a decadent stage, representing a corruption of the previous period. Europeans had applied this same scheme to the classical cultures of the Mediterranean; they reapplied it to Mesoamerica.

The Maya have existed for at least 3,800 years, a time we identify as the start of Preclassic Maya civilization. Again, we see the problematic nomenclature that defines everything in relation to one period, the Classic Period, with names that suggest that the Classic Period is preferable to the incompletely formed Preclassic, or the decadent Postclassic. Archaeologists have long written about what happened to cause such places to be abandoned in the ninth century CE, and this is certainly one of the "collapses" that has made its way into the popular imagination. From at least the early twentieth century, archaeologists talked about the Classic Maya collapse, although many of the details of the timing were wrong.[10] These sites and monuments were the focus of much work by archaeologists and their predecessors since the nineteenth century.

The events we refer to as the Classic Maya collapse became known to a much wider, popular audience because of the supposed apocalyptic prophesy for 2012. The idea that some sort of apocalypse would happen in 2012—spawning all sorts of popular narratives, including movies—was based on the erroneous idea that the Maya prophesied that the end of the world would occur at the end of the thirteenth *baktun*, a unit of time measuring 144,000 days or approximately 394 years. According to one comparison of the Maya calendar with our current Gregorian calendar, this date fell on December 21, 2012. That date passed, and the world did not end. No discernible apocalypse occurred. Anyone familiar with the Maya would contest the assertion that they prophesied the end of the world at all, or that we could reliably correlate their calendar with the Gregorian calendar. The Maya prophesy did not fail to come true; it never existed. While the anticipated events of 2012 did not come to pass, dramatic and profound changes had occurred in the past. The Maya underwent significant changes at various times during their long history. The Classic Maya collapse was not the first or only such event in the region. There were similar events much earlier in their history as well.

Jared Diamond's book *Collapse: How Societies Choose to Fail or Succeed* popularized the Maya situation.[11] He attributed the decline to the elites, the divine kings of the Maya, who created and ignored social problems with their endless demands for warfare and sumptuary goods, including things like an enormous amount of lime plaster for pyramids (featured prominently in the 2006 movie *Apocalypto*), which required an enormous amount of firewood to produce, contributing to environmental degradation.

Environmental degradation, including deforestation, has been posited as a cause for collapse in the Maya area and elsewhere.[12] This explanation has many critics, and other causes have also been suggested for the transformation at the end of the Classic Period, including warfare, disease, and drought.[13] The Maya example provides an insight into how complex past collapses have been. Scholarly opinions about what happened in the ninth century CE run the gamut, from the nature of the transformation to its causes.[14] In the following chapters, I explore potential causes of the Maya collapse, and examine what this event would have looked like for residents of the Maya area at the end of the ninth century CE.

THE WESTERN ROMAN EMPIRE

The second historical example of things falling apart that I explore here is the decline and fall of the Western Roman Empire in the fourth and fifth century CE. The empire spanned the western Mediterranean, including much of Europe west of the Adriatic, and North Africa west of Egypt. I started working in the Mediterranean about ten years ago in Spain, helping to develop an underwater 3-D system. Over the next few years, I continued in Albania, Montenegro, and Croatia, teaching underwater archaeology field schools, helping with scientific diving courses, and participating in the coastal survey for the Illyrian Coastal Exploration Project, focused on the eastern coast of the Adriatic Sea from Croatia to Greece. This was new and exciting to me, and the kinds of shipwrecks we found were vastly different from the ones I knew in Central America and the Caribbean, where I'd worked on shipwrecks dating from the sixteenth to nineteenth century. In the Mediterranean, most of what I documented were shipwrecks between 1,000 and 2,500 years old, full of ceramic jars or amphorae.

In 2016, I was invited to help with a project around the island of Fourni, in Greece, where we found an enormous number of shipwrecks.* The media now call this archipelago the "shipwreck capital of the Mediterranean," which seems accurate.[15] Our small

* The Fourni Archaeological Project was directed by Dr. George Koutsouflakis of the Greek Ephorate of Underwater Antiquities and Dr. Peter Campbell of Cranfield University in England, and I am grateful to have been invited to participate.

team would locate one or two new shipwrecks every day, almost always with the help of local fishermen who knew everything in the water and who would show us the wrecks with which they were familiar. On one occasion, we found a wreck that still had much of the wooden hull left. A nearby amphora, that we can identify as to its age and origin by its shape and design, suggested that this might be a very old wreck. Ultimately, further investigation revealed that this was probably a seventeenth-century wreck. Some of the crew, consisting primarily of Greek archaeologists, were visibly deflated: this was recent, to their way of thinking. To me, on the other hand, it seemed like a very old and well-preserved shipwreck.

After finding so many shipwrecks, enthusiasm for finding a new wreck had died down a bit. When the wreck was relatively old, or from time periods about which we knew little, the enthusiasm was back. Byzantine wrecks were better than younger medieval wrecks, but not by much. The slightly older Late Roman wrecks from the fourth or fifth century CE are of particular interest if we are seeking to understand how societies decline. During this period, Rome was declining. At the end of it, we have the fall of the Western Roman Empire. The late Roman period largely constituted the "decline" in the oft-cited "decline and fall of the Roman Empire."

If the Maya collapse is the best-known example of such a thing in the Americas, the decline and fall of the Western Roman Empire must be the most discussed decline globally. As the Maya collapse is sometimes couched in environmental terms, the collapse of the Roman Empire is seen as one resulting from social and political issues. The general story of the development and collapse of the Roman Empire is well documented, most famously with the publication in 1776 of Edward Gibbon's six-volume *History of the Decline and Fall of the Roman Empire* and in innumerable publications since.[16] The Roman Republic fell in 27 BCE, replaced by the Roman Empire, which endured until the fifth century CE, with severe challenges occurring throughout its history, worsening in the late fourth century. The decline and fall of the Western Roman Empire, in the most general sense, results from a number of issues building up over two centuries or so, culminating in the deposing of the last emperor, Romulus Augustulus, in the year 476 CE, or, for some historians who consider him illegitimate, Julius Nepos in the year 475 CE. This was a long decline and fall.[17]

EASTERN NORTH AMERICA

The third example I explore comes from my home state of Kentucky and represents my very first archaeological experiences. I first excavated there during my freshman year in college, and I worked in Kentucky for several years before turning my attention to Central America in graduate school. I never quit doing archaeology in Kentucky, though, and I now have an underwater archaeology research project in the waterways that traverse the commonwealth. The archaeology here is interesting, this is my home, and there is a community of excellent archaeologists here.

One of my favorite memories excavating in Kentucky happened just after I graduated from college. I was waiting on a paying job to start later in the summer, but I needed work right away. I called one of my friends and mentors, archaeologist Cecil Ison of the U.S. Forest Service. I asked if he had any projects starting soon, and if so, if he needed another crewmember for archaeological work. I had just returned from a short trip to Europe, my first trip out of the country, and I had no money and no place of my own to live. Cecil had a project starting the next day, in fact, but had only a bare-bones budget, and no money for an extra crewmember. He described the project, an excavation at a rockshelter in eastern Kentucky that had some very old deposits. I knew the area, and it sounded interesting. I asked if I could work anyway, in exchange for room and board in the state park lodge where they were staying for the duration of the project. He agreed. Later that afternoon, I stood on the sidewalk in downtown Lexington, on the courthouse square, with my backpack of tools and a duffel bag of clothes, waiting for a truck to take me two hours east.

I squeezed into the very back of the rockshelter. I lay on my stomach and tried to reach down to the bottom of the small excavation unit. I had been working with the Forest Service archaeologists for about a month by that time. We were excavating a site high on a mountainside in eastern Kentucky, in a small rockshelter formed by an overhanging cliff. Folks looking for artifacts had looted most rockshelters, and often the only intact parts are in the inaccessible corners. I was working way back in one of those corners. I could not stand up, and so I scooped dirt out with my trowel into a bucket, then wiggled out to the front of the shelter and passed the dirt through quarter-inch screen to recover small artifacts. I was down so deep I could barely reach the floor of my excavation, but the artifacts kept coming. Near the bottom, I found a large chunk of charcoal, which I put in aluminum foil and labeled. This would be sent off for carbon-14 dating. A little lower down, the artifacts disappeared and I moved on to excavate somewhere else. A year later, I heard that the oldest date for an archaeological site in Kentucky had come from that rockshelter—over 13,000 years old. I have often wondered if it was from that piece of charcoal.

Fast-forward 13,000 years, and there are few Native American people living in that region, having been forcibly removed by the federal government in the early nineteenth century. Many people in Kentucky claim Native American ancestry now that it is fashionable, but other than these purported relationships, there are very few of the original inhabitants of the Commonwealth, and the places in which people lived for millennia are gone except for traces. We have cultivated ways to explain away this situation. In Kentucky, we have the myth of the "dark and bloody ground," which misinforms us that this area was not the permanent home of Native Americans when Europeans arrived. Rather, the myth suggests, it was a contested area, bloodied by warfare and not occupied permanently by any group. Other stories would have us believe that Kentucky was a hunting ground prior to the arrival of James Harrod, Daniel Boone, and the rest of the colonizers who established settlements in this area. None of that is entirely true, and we have known this for a long time. This story endures, however, and it came in handy when occupying the land and subsequently explaining to ourselves how it is that none of the original inhabitants of the area remain. This kind of convenient narrative is common throughout the Americas.

I grew up hearing those stories. So pervasive were they that when I became interested in archaeology, I thought of it as something you had to do somewhere else. Peru or Egypt, perhaps, but not in Kentucky. I did not know the 15,000 years or more of human history in my home state. I did not know what fantastic archaeological sites we have in Kentucky, nor what accomplished archaeologists we have. I found all of that in college when I had a chance to work with some of them, excavating at sites around Kentucky. I met many of those archaeologists on my first archaeological excavation when I was seventeen years old. I volunteered to help on a weekend excavation at the Snag Creek site, a late-prehistoric village site overlooking the Ohio River. You get there on Mary Ingles Highway, named for the "first white woman" in Kentucky. We do not know the original name of the village, or the name of anybody who lived there.

Like many people in Kentucky, I had grown up finding the odd arrowhead or spear point in a plowed field or rockshelter. That weekend at Snag Creek, I learned how to identify and excavate the ephemeral remains of houses and to piece together some image of the past from what we found. We found postholes, defining the edges of houses. We found trash pits, originally for storage but later filled with ash and garbage when they were no longer needed.

The site we investigated that weekend was around 500 to 600 years old—not particularly old in the scheme of things. I have been in a pub in England older than that. I stood at the gates of a farm in Iceland that had the same name as it did in a Viking saga written nearly a thousand years ago. In Kentucky, we do not even know what language the indigenous people spoke six hundred years ago, or what they called themselves.

Archaeologists divide the past into different periods, based on differences we see in the artifacts, or material culture, of a particular area. We also identify various archaeological "cultures," again based on the materials left behind, often referred to by the name of the site where archaeologists identified the culture. Sometimes we can associate those archaeological cultures with historic or modern groups, but often we cannot. Archaeologists' categories of time or cultural affiliation may or may not have been important categories for the people who lived in the past. Most of the time periods are long enough and the change gradual enough that nobody living through it would think of time in the way we divide it. From examining contemporary people, we can see the limitations and problems with basing cultural affiliation on material culture. Neighbors, like Hondurans and Salvadorans, are distinct, from variations in vocabulary and accent to important historical differences that make life unique in those two modern nation-states. The vast majority of their material culture, however, is identical. In other instances, differences in material culture obscure similarities elsewhere.

Archaeologists call the period just before the arrival of the Europeans the Late Prehistoric Period, followed by the Contact Period, when the groups met. We use the name Fort Ancient to refer to one of the archaeological cultures we have identified, named after a well-known site in Ohio. The Snag Creek site was a Fort Ancient village. Over the next decade, I worked on and off at several Fort Ancient sites, as part of the efforts of archaeologists to understand life just before everything changed.

Humans like us, modern humans (*Homo sapiens sapiens*), evolved in Africa around 300,000 years ago. After a few fits and starts, some of those people moved out of Africa and across the globe at least 75,000 years ago. One of the last places we moved to was the Americas. Humans entered the Americas from Asia somewhere around 20,000 years ago, and maybe earlier. Blocked from traveling overland by extensive ice sheets in what is now Canada, the populations at first would have moved south by boat, living on the ice, living off the sea. About 13,000 years ago, an ice-free corridor opened up east of the Canadian Rockies that allowed overland passage from Beringia (an area stretching from Russia, across the Bering Strait, through Alaska, to Canada) to the rest of the Americas. By at least 15,000 years ago, people were living all across both North and South America. Every year we find more evidence of earlier occupations, and I would not be surprised if we eventually discover that this date was more like 20,000 or 25,000 years ago. We know there were people here in Kentucky by at least 13,000 years ago, as suggested by carbon-14 dating of the charcoal from the rockshelter.

At that time, people were foragers, or hunters and gatherers. They did not farm, but it only makes sense that they knew all about plants, as they gathered them and probably manipulated them to some degree. When I traveled through the rainforest in Honduras, we planted fruit trees or watermelons that might be ready the next time we came through. It stands to reason that people did similar things throughout history, and archaeologists have found evidence of people manipulating plants long before agriculture became the dominant strategy.[18] As time went by, the population grew until they could not feed everybody through foraging. With that new need, they became farmers. This did not happen overnight, and I suspect it was not the result of a "discovery." It was something they had to do, amplifying something they were already doing. Foraging typically takes much less time than agriculture to secure the food you need when populations are low enough. When population density increases, people switch to agriculture. In parts of the Middle East, including Syria and Iraq, people became agriculturalists around 12,000 years ago. In areas such as Peru, the Amazon, and central Mexico, people switched to agriculture by 6,000 years ago. In other areas like eastern North America, where resources were plentiful and population was still relatively low, agriculture began about 3,000 years ago.

By the time we have the first written records of life in the area, Kentucky was home to several different groups, including at least Cherokee, Shawnee, Yuchi, and Chickasaw, and probably others.[19] What we do not know is what things looked like just before the Europeans arrived, because of the transformative, often catastrophic events that accompanied their arrival. For a long time, the stories told in the Americas about Native Americans prior to the arrival of the Europeans were the stories that worked for the Europeans. These narratives minimized the catastrophic impact of the encounter or made it easier to justify the subsequent actions by colonizers. For instance, until the mid-twentieth century, scholars routinely underestimated the number of people living in the Americas prior to the arrival of the Europeans. Cultural geographer William Denevan's estimate of around 54 million changed all that.[20] His has long been the gold-standard estimate. Recent studies suggest slightly more, at around 60 million.[21] Earlier estimates radically underestimated the number, with 10 million or so being the typical estimate in the late nineteenth century. Assuming a low population makes it possible to ignore the huge demographic collapses that occurred, and to suggest that the area was largely vacant and that occupying the land was justified.

The hunting-ground myth in Kentucky is another, local, way to underestimate the number of people here prior to the arrival of colonizers. One of the first archaeologists with whom I worked, Dr. Gwynn Henderson of the Kentucky Archaeological Survey, has written about and worked to debunk this story.[22] "The most enduring fallacy about Kentucky's indigenous inhabitants—the myth of the Dark and Bloody Ground—concerns how these peoples used the land," Henderson recounted in a recent conversation. "This legend would have us believe that Indians never lived permanently anywhere in Kentucky, but only hunted and fought over it. The myth has been and *continues to be perpetuated* in children's books, in scholarly books and journals, in histories, and in magazines. It persists despite the fact that *Kentucky* is simply a geographic construct," with no meaning until 1792, and "despite the fact that no such notion exists for the surrounding geographic constructs of Ohio, Indiana, Illinois, Missouri, and Tennessee." In Kentucky, because of warfare between groups, there was a moment when settlement tended to be consolidated behind the front lines and fewer villages were located in a certain part of the state. At no time, however, was the area completely depopulated.

In the Americas, we find the closest thing to the kind of event depicted in some of the fictional apocalyptic narratives that are ubiquitous today, in the sense that a large percentage of the hemispheric population died. How this happened, and how rapidly, varied across the continents. Native American groups persisted, however. Although these narratives of destruction at the hands of European disease and oppression can create the impression that these groups did not somehow persevere, they did. The Native American cultures that underwent profound transformations with the arrival of Europeans are as important, interesting, and vital as what existed before. The disruption that occurred when Europeans arrived in the Americas did not destroy the cultures that existed here, or somehow result in cultural extinction. Describing what happened as "destruction" or "extinction" can allow a justification for the types of marginalization that has proved so damaging in the past. Many archaeologists now understand this, and reject these "terminal narratives" that purport to explain the demise of Native American groups. In some ways, this is a one-two punch. First, we create narratives that downplay the destructive nature of the interaction.[23] Then we suggest that authentic Native American cultures have been damaged beyond recognition. The lesson communicated is that Native Americans no longer exist, for all intents and purposes. That is not true.

The catastrophe that resulted from the outbreak of disease following the arrival of European colonizers in the Americas was unprecedented. The warfare and enslavement that followed exacerbated this tragedy. Historian Gerald Horne describes two apocalypses; the first related to disease, and the second resulting from the horrors of settler colonialism.[24] We see this kind of colonialism in North America, where colonists come to settle an area, not merely to exploit it for resources. This resulted in the continued genocide of Native Americans, partly by disease but also through enslavement, displacement, and murder. In addition to that, settler colonialism involved the enslavement of millions of Africans. In the following chapters, I examine the events in eastern North America in the sixteenth and seventeenth centuries. I begin by exploring the causes of these past catastrophes, seeking an understanding of why things fall apart. This allows us, later, to compare our contemporary ideas of how and why things fall apart, and to look into the future to the point when this process could begin again.

CHAPTER 2

WHY THINGS FALL APART

My wife is a fashion designer. She calls herself a dressmaker, because she feels "designer" sounds pretentious and to show solidarity with the many people whom she worked with in sweatshops in Honduras when she was a teenager. She wonders how many designers or artists might have emerged from those sewing rooms if people's creativity had been fostered, rather than smothered by the crushing demands of providing for their families. We now live in Lexington, Kentucky, and she is able to work creatively. She started a nonprofit, the Lexington Fashion Collaborative, to encourage and enable an outlet for the type of creativity that was stifled in her coworkers in the clothing factories. In March of 2020, we were a week out from an elaborate two-day event that she had organized over the previous year, featuring a dozen designers from all over the southeast United States, scores of models, and several groups showcasing a diversity of cultural performances. As she was in the midst of the last-minute preparations for the event, I drove the hour and a half to Louisville to attend the Louisville Survival Expo and Gun Show, as part of the research for this book.

I had heard about Covid-19 by then, as we all had, but I had no sense that it could seriously affect our lives. Even as I was working on a project about apocalyptic events, I did not anticipate the scale of the catastrophe to come. I had read or watched dozens of fictional narratives about catastrophic outbreaks by that point, but they tend to focus on much deadlier diseases, or they are zombie fantasies. I did not see the real-life version on the horizon. That changed as I sat at the expo with Adam Nemett, author of the postapocalyptic novel *We Can Save Us All.*[1] We listened to organizer and featured speaker Bob Gaskin talk about current threats to our way of life. He talked for forty-five minutes about the virus, and the effects it would have when schools closed and supply chains broke down. I listened skeptically, as I had listened in the past to similar presentations about the threat of an electromagnetic pulse or the always-imminent economic collapse due to our abandonment of the gold standard. Gaskin was a compelling speaker, however, both rhetorically and because of an ability to articulate the potential effects on the systems and structures that undergird our modern society. Later, debriefing with Adam about the presentation, I remained skeptical of what we had heard. Schools were not going to close, I thought. This will blow over. I was wrong.[2]

Later that week, my wife made the tough decision to cancel her event. In the course of a few days, everything had changed. My university extended spring break for a week, and school cancellations seemed inevitable. Two days before her event, the governor advised that we cancel or postpone all community gatherings. Everything ground to a halt. At that moment, canceling that event seemed like a big deal. Later, in retrospect, it was a mere inconvenience.

The pandemic was not an apocalyptic event, but it was tragic and it changed everything. We are familiar with the lives lost, and livelihoods. Everything shut down. Some closures were temporary, but many were for good, especially among the small local businesses that we took such pride in as a community. Looking back, I see how long it took some of us to recognize what was happening. Archaeologists and historians have long recognized that people do not realize a collapse is upon them until it is far advanced. In this case, the scope and impact of the pandemic was not immediately obvious, and most of us did not understand the scale of it for weeks or even months.

The pandemic might not be an event on the scale of what happened in the archaeological examples I've mentioned, but it can be illustrative in thinking about the cause of catastrophe. Any pandemic has a pathogen as a cause, but that is not the sum total of what creates the catastrophe. Our highly interconnected world contributes to the impact. Air travel and cruise ships were at the forefront of the spread of this virus. Our system of global supply chains was more vulnerable to disruption than a local supply chain. The virtually nonexistent response by the U.S. federal government allowed the pandemic to spread faster and kill more than it should have. The anti-scientific and anti-intellectual tendencies of a segment of the population contributed to the catastrophe, too, as did the politicization of protective measures like masking, social distancing, and shutdowns. When we look at the catastrophe that was the Covid-19 pandemic, the virus itself is only part of the story.

Discerning the cause of past apocalyptic events involves identifying all of the elements that contributed to the collapse, and understanding their relationships. Trying to isolate a singular cause (or set of causes) that precipitated the catastrophe would not be sufficient to explain what happens next. The term *proximate cause* refers to the forces that set off the catastrophic event in the first place. The proximate or original cause may not be as evident in some historical events as it is during a pandemic. Collapses are multi-causal and rarely complete. Our focus on the discontinuity, the changes, can obscure the fact that peoples and cultures persist even when specific political and social structures collapse. In examining the causes of collapses, we can see the complex relationship between those causes and the systems that ultimately change. We also see that, despite the discontinuity, there is always continuity. People and structures survive, and communities re-form. Our fantasies of individualist survival are at odds with the reality that communities, groups beyond our families, will always be how we survive. The difficulty of determining proximate causes is evident in an examination of the causes of three ancient transformations discussed here.

THE CLASSIC PERIOD MAYA

As discussed earlier, archaeologists see a decline during the ninth century CE in the southern Maya lowlands, but the area is large and not homogenous, and various times and places showed different kinds of change. Not everybody agrees on how rapidly the changes occurred, how significant they were, and there is even less consensus on why.[3]

To understand the cause of a collapse, it is important to know whether an entire area experienced a homogeneous transformation. If so, it might be that similar forces were at work across the entire area. If not, positing a singular cause is less compelling. The short answer is that no, we do not see a homogeneous event in this case. Human remains, paleopathology (looking at disease and injuries), and paleodiet analysis show local variation and diversity around the Maya lowlands in terms of disease load and diet during the time of the Classic Maya collapse, suggesting there was no large-scale, homogenous set of conditions or causes.[4] We might expect to see similarities in nutritional stress and disease across the region if there were a single, large-scale environmental phenomenon. On the other hand, I can imagine that a widespread phenomenon, environmental or otherwise, could play out differently in various areas because of other factors that determined the effects in each local area. We can see this in modern situations, where a drought causes famine in one place but not in another, and we see that the catastrophes that result from natural phenomena have as much to do with political realities and food distribution strategies as they do weather.

Some archaeologists suggest that the end of the Classic Period in the Maya Lowlands took too long (about 200 years, from 750 CE to 950 CE or later) to be categorized as a collapse.[5] They note the ways in which the changes evident in the Terminal Classic—the very end of the period when the collapse occurred—varied across the region. The end of the eighth century was a time of warfare in some areas, and certain cities seemed to decline at the same time, but there was heterogeneity throughout the Maya Lowlands. New activity, such as the search for new trade routes, the movement of population, and the introduction of new imported goods continued in some parts of the area, while other places seemed to shut down.[6]

Archaeologists know that similar collapses happened earlier in the same area but played out differently. One such event occurred several hundred years before the Classic Maya collapse, at the end of the Preclassic Period. That event resulted in a very different political situation from the later one.[7] There is evidence of drought during both of these periods, with the drought being worse to the south, and this corresponds to a greater decline seen throughout the southern part of the region. The different results of the two droughts could have to do with the types of coping strategies used to deal with severe drought, which failed to work in the late Classic period but might have worked during the earlier drought.[8] One strategy might have been a shift away from a broad agricultural regimen to one that focused more intensively on corn.[9] The more complex social and political systems in place by 750 CE made adaptation to the drought less successful.

Some archaeologists exploring this earlier collapse also see evidence that a foreign power intervened, either directly or indirectly, and this interaction at a time of profound change resulted in the emergence of a powerful dynasty that ruled Maya cities like Ceibal during the Classic Period.[10] The decline at the end of the Classic Period did not result in the replacement of one regime with another. After the Classic Period decline, significant activity at the site ceased. This difference in the two declines, both of which might have been precipitated, in part, by drought, suggests that the way things play out is dependent on more than the proximate cause. It is the collapse of the complex systems, and how they are reconstituted, that shapes the collapse or transformation.

Some of the research described above was conducted by archaeologist Takeshi Inomata of the University of Arizona. He works directly with a lot of these issues, including drought, at the site of Ceibal in Guatemala, among other places. I spoke with him about drought and other issues related to the Classic Maya. We talked about whether the concept of collapse is an accurate way to describe what happened there. He suggests it might be premature to reject the concept of collapse completely. "I use the term 'collapse,'" he told me. "In some areas, we see a population decline of 90 percent, or complete abandonment. I think that in those areas there was political disintegration and a major social change that affected all the population. But we need to use the term cautiously. In the past, it was used to allude to the image of a 'lost civilization,' and the public may still get that impression. Instead of trying to characterize the complex process in a simple term, whether 'collapse' or its replacement words, we need to analyze and describe it carefully." Having been enmeshed in ridiculous controversies over the claims of discovery of lost cities (never lost to local residents, only to the interloping "discoverers"), I appreciated his caveat that we not conflate the collapse of a particular political system with the mysterious disappearance of some lost city, awaiting discovery in the rainforest. However, he suggests, the speed and scope of the transformation warrants the use of the term in some instances. Inomata also notes that the Maya area was not homogenous enough that one could talk about "the collapse" as a single phenomenon. Significant differences existed between the processes under way in the southern lowlands (the rainforest in Guatemala, Belize, and Honduras) and those in the northern areas like the Yucatán. The concept of a collapse works for some places at some times.

Inomata's work at Ceibal sometimes focuses on the evidence for drought. I asked his opinion on the question of whether changes at the end of the Classic Period were largely the result of environmental issues, such as the drought for which he has found evidence, and whether he would find it compelling to label it a collapse. He noted, as others have, that this is still an issue that needs more investigation, and a clear answer has not emerged. He did think environmental issues were important in the early stages of the decline. "A possible scenario is that in an early stage the combination of social and environmental problems may have been important," he notes. "The social problems include competition, warfare, and rapid population increase. The environmental ones may be deforestation and soil erosion. Evidence for droughts in the early stage is weak. There was then an effort to reestablish various polities, but then major droughts may have come. If so, droughts may have been the last blow to the society which had already been weakened by other problems." As we continue to see, these transformations are complex and multicausal, even if a proximate cause can be established. In Inomata's view, the proximate causes were not the same causes that drove the collapse in its latter stages. Drought, for instance, did not come into play until the collapse was already under way.

The data suggest that environmental factors were significant in some places at some times, but establishing a proximate cause is difficult. Some compelling research has found evidence of drought at the end of the Classic Period. This evidence for significant drought comes from several sources. One project found data suggesting significant drought from 700 to 1135 CE in Western Belize, in the Maya Lowlands from stalagmites in caves.[11] There is a correlation between "dwindling water supplies" and the decline seen in the ninth and tenth centuries CE. Some archaeologists think the role of drought has been overestimated in some areas, and that drought is insufficient to explain the nature and variability of the changes seen at the end of the Classic Period.[12] Positing drought as *the* cause of the collapse, rather than one of many factors that contributed to profound societal changes, has fallen out of favor with many scholars working in the region, with entire edited volumes dedicated to this issue.[13]

From all of the variations throughout the Maya area and nearby regions, it seems that, ultimately, a multicausal cascade of events shape the transformations, including those labeled collapses. One of the common conclusions that archaeologists draw when they look at dramatic changes in the past, some of which we've labeled collapses, is that they nearly always have multiple causes. There are some exceptions, such as volcanic eruptions. In most cases, however, there might be a proximate cause—that is, something that starts the whole process of collapse—but ultimately it is a series of events that actually destroy a particular system. In the Maya area, for instance, we see drought, deforestation, warfare, and the increasing cost of maintaining the elites as possible causes for the changes that took place during the ninth century all over parts of the southern Maya world. Looking at the region as a whole, we see that there are differences within the region, and the simplistic conclusions of the past are now being rejected. So while environment played a role in the Maya collapse, it would be reductive and inaccurate to think of this as an explicitly environmental event. Regardless of the cause or causes that sets the whole thing in motion, the collapse of complex human systems are always complex as well.

Archaeologists tend to see our own problems in the past, and our interpretations reflect our contemporary concerns.[14] We overlay, or "retrofit," to use McAnany's term, our concerns onto the past, and see causes that we worry about in the present day. I see the attraction of retrofitting our concerns onto past situations, as it allows the past to serve as a "cautionary tale," and archaeology as relevant to the modern world as a sort of early warning system. Archaeologists identify environmental problems as we confront climate change. Warfare is another cause often cited for collapse, either as a proximate cause or in conjunction with other problems. "Warfare and environmental destruction, those seem to be the most popular explanations right now," says McAnany. We talked about the Covid-19 pandemic. "I predict there's going to be some kind of resurgence in theories of Maya collapse as a result of pandemics," she said.

One reason why we (archaeologists) have not resolved some of these basic issues relates to the type of evidence we use. Often, we do not have direct evidence of something, like warfare. We typically do not find the combatants, nor defensive structures (with a few exceptions), nor caches of weapons. Rather, we find indirect evidence, like patterns of abandonment among settlements, and in the case of the Maya, textual evidence in the form of hieroglyphic texts on monuments and buildings. Interpreting this indirect evidence can be difficult and contentious.

Sometimes we do not have data from multiple time periods that allow comparisons to be made. In order to posit something like an increase in warfare, you need to have a baseline to measure against. McAnany made this point in our conversation:

> I think there are some problems . . . making a statement like "warfare really accelerated in the late Classic." People say that all the time. And what do they base it on? They base it on the hieroglyphic texts that describe some kind of martial event

> between two places. The only problem with the statement is that we don't have comparable evidence from the early Classic or from the Postclassic. And so you're talking about something becoming more frequent and more intense, but compared to what? Are we assuming that things were just very calm and peaceful running up to the end of the early Classic and then you have lots of martial activity in the late Classic and then none again in the Postclassic? We know the Postclassic to have been a time of active conflict. These were polities that were structured . . . kind of like medieval European kingdoms. So warfare was a fact of life there as well.

Most archaeologists believe that any kind of collapse ultimately results from the disruption of the complex systems that make up a society, not directly from the proximate causes that set the disruption in motion. McAnany related this idea in our conversation, focusing on the disruption of the alliances that existed in the Maya world at the end of the Classic Period. "This whole alliance structure was, I think, probably key to the whole thing. And that broke down for some reason, and I think that's probably what began to make the place unlivable." Even the compelling evidence for a drought needs to be understood as something that begins to break down the complex systems needed to keep everything going. She emphasizes that a drought affects a system, and we need to examine that system.

"If you think about a time when there was no United Nations to deliver tons of grain to an area when people were starving, you realize that food and provisioning these places probably was always an issue. And people did go hungry. Once again, I think [we have to look] back to this alliance structure that we see politically." The structures that fail when stressed create the crises.

Drought data is ambiguous, but rainfall did decrease unevenly across the region between AD 750 and 1050, although drought does not explain the complexities of the archaeological data.[15]

The evidence of drought is still spotty, however, and some areas declined in one period and then recovered in the next, in such a way that is inconsistent with what we might expect from widespread, sustained drought. In any case, in McAnany's view, drought would be one potential influence on a political transformation, rather than being an explanation for significant change in and of itself.

One issue to consider when looking for the cause of a past catastrophe is how changes, especially crises, would put pressure on existing political systems. The fragility of these political systems might surprise us. McAnany returned to this theme in our conversation: "Political systems are much more delicate than people want to consider. They are inherently fragile. I think we're realizing with our own system, which we thought was such a robust democracy, how very fragile that is. I think that political systems come and go for that reason a lot more quickly than, say, religious systems or social systems, which tend to be much more durable." In the months after our conversation, that political fragility became more apparent than ever.

We see the fragility in the systems that collapsed; they did collapse, after all. We also see a great deal of resilience in those moments of great change. The nature and degree of resilience varies with the particular event, just like everything else does. For instance, Inomata sees evidence of great resilience after the Maya decline at the end of the Preclassic period, but not the same type of resilience after the Classic Maya collapse. "At the end of the Preclassic, after the collapse of [large sites, or centers, like] El Mirador, et cetera, other centers like Tikal emerged. It certainly represents resiliency." That same resiliency was not evident in the second collapse. He continues: "This same area showed a drastic depopulation, without much recovery in many parts, at the end of the Classic. We see much more continuity in the north, but we do not understand why the southern lowlands were hit so hard and their recovery was so weak."

Even though people are resilient and bounce back, this should not obscure the fact that the decline might have been profound. Inomata warned against a tendency to minimize the collapse because it was long, complex, and multicausal. "It was certainly a long and complex process," he states, "but some people overemphasize its diversity and gradual nature." Part of this tendency might result from our chronologies, which are incomplete or lack the resolution to narrow down exactly when something happened, and what else was happening at the same time.

Clearly, there is no consensus on exactly what precipitated the events at the end of the Classic Period in the Maya lowlands. Nor is there consensus regarding how the decline happened, or whether it could be described as a "collapse," or whether the events were widespread and homogeneous. We do see a consensus, or as close as archaeologists get to one, around the idea that the Classic Maya collapse was a long and complex series of events, with causal events that might have shifted over time and space but that triggered the breakdown of the kind of complex social, political, and economic systems that were in place at that time. Any of these could trigger another reaction that would perpetuate the process until change was profound.[16]

Even without a consensus, we see that the real, significant changes start with the collapse of the complex systems that keep everything going. In imagining our own future, three things seem clear from this archaeological example: First, our systems are fragile, and we might not be able to predict what proximate cause or causes set in motion an event that leads to their collapse. We need to safeguard these systems if they are ones we want to preserve. Second, profound change can begin slowly, and can have a long duration. Our image of a sudden collapse—here today and gone tomorrow—does not resemble what happened with the Maya. Even though the progression of the decline may not be linear or steady, and even if it suddenly gets worse, the Maya example suggests that declines are a long time in the making. That means, in the best case, we might be able to identify a process that has already started, like climate change, that has the potential to create systemic failure if there is no course correction.[17]

The Classic Maya collapse is one of the more contentious historical collapses in terms of why it happened, what it looked like, and what we should call it. Even the most general statements can be challenged by some legitimate and compelling interpretations. My purpose here, though, is to look at the range of what happened in the past and compare it with the ways in which we envision a future catastrophe. All of the varied conclusions that researchers might reasonably come to with regards to the Classic Maya serve to illustrate the ways in which our fantasies of the next apocalypse do not resemble anything from humanity's long history, and point away from an individualistic approach toward one that foregrounds community and cooperation.

THE WESTERN ROMAN EMPIRE

I can think of no event analyzed more by scholars than the decline and fall of the Western Roman Empire, nor an event for which so many causes have been proposed. Unlike the other examples I examine, the decline of the Western Roman Empire has extensive textual data in addition to the archaeological and has been the subject of analyses innumerable times over the intervening centuries, beginning with analyses that were contemporary with the actual events.

In order to hear the current state of thinking about the fall of the Western Roman Empire, I spoke to several archaeologists who work in the area, on issues directly or tangentially related to the fall. I spoke first with Riccardo Montalbano, an archaeologist and researcher at the University of Pisa, a native of the region who has studied these issues directly. He has worked around Europe, including Rome and Venice in Italy, and in England and Germany. Dr. Montalbano studies ancient urbanism, landscape archaeology, and new technologies for territorial analysis, with a special focus on the Mediterranean during the Roman Age.

"It is not easy to deal with such a complex issue in a few lines, when we think how long the historiographic debate is," he tells me. "Although in the past there have been attempts to isolate specific elements [as causes, including the embrace of Christianity, the economy, or over-investment in the military], the truth is that the process that led to the fall of the empire is not linear, and it is quite impossible to establish a precise 'starting point' for its beginning. It was certainly a convergence of causes, both internal and external. In any case, it is worth remembering that the crisis factors had different impacts on the two halves of the empire: the consequences were most serious in the western part, while the Roman east faced better some of the dynamics of the ongoing destructuring." Again we see multiple causes and a variable response to the same stimulus shaping the decline, just as we saw with the Classic Maya collapse.

Archaeologists must look at the details of the context in which the decline occurred. Some crises began long before they manifested as a crisis. Overinvestment in the military in one moment sets off a chain of events that culminates decades later in a decline of the middle class. This describes what was happening in the middle of the third century CE. The crises became more numerous and set off a chain reaction, or a series of chain reactions.

Montalbano explains some of the salient forces working to break apart the Western Roman Empire. By the middle of the third century CE, "the borders are now constantly threatened by the barbarian populations (the main external cause) and the growing military needs to cope with the threat, which involves excessive expenses, resulting in financial crises." The Roman state adopts increasingly burdensome taxation to fund the military. Repeated barbarian invasions fragmented political and trade systems, disrupting pan-Mediterranean trade, which led to a decline in living conditions and overall health. Wealth inequities and a decline in rural populations and agricultural productivity compounded these problems by the fifth century CE.

In archaeology, the sources of our data are obviously important, and can influence our interpretations. For the Western Roman Empire, both literary and epigraphic (from inscriptions) sources exist, but these texts do not exhaustively deal with the variety of crises that the empire faced. Also, unlike archaeological data, literary sources might be produced with personal or propagandistic agendas. Archaeologists have personal and propagandistic agendas as well, but archaeological data itself is not produced with any particular intent.

Despite the availability of textual evidence, archaeology contributes much to this reconstruction. In fact, the historical summaries derived from textual sources are running up against an increasing amount of archaeological data, and this does not always agree with the historians' reconstructions of events. Montalbano pointed out another aspect of the historical reconstruction of the decline: He noted that archaeological research today considers previously neglected geographical spaces, such as Eastern Europe. Because of this extended geographical context, we should avoid treating the theme of the fall of the empire in a monolithic way, either in terms of chronology or in terms of the nature of the changes.

I also spoke with Dr. Mantha Zarmakoupi, now a professor of Roman architecture at the University of Pennsylvania. I worked with her in Fourni, Greece, on an underwater archaeology project, and I knew she had done extensive work in the eastern Mediterranean, with a focus on the Greek island of Delos.[18] While her main interest was never the collapse of the Roman Empire, she did see how Delos recovered from more than one episode of destruction. Importantly, she identified places in the textual accounts that disagreed with archaeological data. For instance, during an episode of destruction in the first century BCE, archaeological evidence confirms the textual suggestions of an intentional destruction of the settlement, including burned houses and closed streets. However, archaeological evidence suggests a recovery into the late Roman period, with the construction of aqueducts, baths, and other structures long after the island was described in texts as abandoned or deserted. Zarmakoupi reminds us not to lose sight of the complexity of any decline or collapse, and that one perspective on an event might not capture the totality of the experience. She notes that while Pompeii, Herculaneum, and other towns were destroyed by the eruption of Vesuvius in 79 CE, communities on the other side of Mount Vesuvius survived and thrived.

The fall of Rome has multiple causes, and it is hard to suggest a proximate cause. Early historians, including Gibbon, suggested that the acceptance of Christianity was a cause of the decline.[19] There was the division of the empire into two parts in the third century; there were invasions by barbarians and other groups; there was an overexpansion and overspending on the military; there was the dependency on slavery; and there was generalized corruption and economic problems. Some scholars suggest that the empire never recovered from a pandemic, the Antonine plague, from 165 to 180 CE, which was probably smallpox or measles.[20] Historian Edward Watts suggests that disregard for political norms was the beginning of the end for Rome's republic.[21] Watts's conclusions put me in mind of McAnany's comment that scholars often notice the problems in the past that resonate in the present.

Again, like with the Classic Maya collapse, multiple causes cascade and reverberate off one another, creating a growing litany of disasters that affects the region in a non-homogeneous way over a very long period of time. The beginning of the end occurs even before the idea of an "end" is entertained.

EASTERN NORTH AMERICA

As far as we know, the European presence in the Americas begins in the early tenth century, with Norse forays into Canada, but these visits were small and short-lived. European colonization of the Americas began in earnest in October of 1492, when Christopher Columbus landed his three ships in the Bahamas. For the next fifteen years, European presence in the Americas was largely limited to the Caribbean, or with only small forays onto the mainland, such as Columbus's landfall on the mainland of Honduras in 1504 during his fourth voyage. More than a decade passed before the first large-scale movement of Europeans onto the mainland occurred.

In North America, Europeans did not arrive until 1513, with Ponce de Leon in Florida.* Over the next thirty years, there were several expeditions into North America, including those of Verrazano, Gomez, Cabeza de Vaca, Coronado, and Cartier. For the southeast area of the present-day United States, the most important of these early entradas (the exploratory trips) was that of Hernando de Soto in 1539. De Soto landed near Tampa in 1539 and proceeded north up through Alabama and Georgia, into Tennessee, and then west to the Mississippi River, where he died.[22] The rest of his crew continued on, and we have three first-hand accounts of the trip, and another that was written decades later. De Soto had over six hundred men, twenty horses, and two hundred pigs traveling with him. In his communications with the powers that be in Spain, De Soto described his journey as one through agricultural fields, orchards, and large villages. He described the large and powerful chiefdom of Coosa, located in what is now in northern Georgia and adjacent parts of Tennessee and Alabama.[23]

Twenty years after De Soto's journey there was an expedition by Tristán de Luna y Arellano, who anchored in Pensacola Bay, Florida, in 1559, and set up a settlement called Santa Maria de Ochuse, recently rediscovered in the modern city of Pensacola.[24] A decade after Luna, Juan Pardo made an entrada. No consensus

* John Cabot explored the coast of Newfoundland in 1497, and João Fernandes Lavrador mapped coastal regions of Labrador in 1499 but did not establish settlements.

exists among archaeologists and historians regarding the great differences in the accounts of 1541 versus those from 1560 or 1568. Until the 1990s, archaeologists and historians tended to interpret the difference as evidence that a major decline had occurred in the region as disease followed in De Soto's wake. Now many historians see no evidence for a decline in the accounts. Instead, it seems that De Soto's accounts since his entrada set up such lofty expectations that the reality of the region did not live up to, thereby creating the illusion of a decline. When Luna and Pardo entered, they might have seen something very similar to what existed in 1541, but nothing like the inflated narratives that had become popular in the interim.[25]

A number of diseases devastated indigenous populations in the Americas, but archaeologists and historians continue to refine our understanding of how this happened. We know that ultimately, after a century or so, the population of Native Americans declined by around 90 percent. Much of this was a direct result of mortality from a number of diseases they had never before encountered, and for which they had developed no immunity. Popular imagination has smallpox as the principal pathogen, but there were many others, including measles, influenza, typhus, and chickenpox. Some diseases, like chickenpox, if acquired during childhood are not particularly dangerous. Adults, on the other hand, have a much different experience, and it can be incredibly dangerous and deadly. Not only did the diseases kill people directly, but also the large number of sick people made it hard to provide basic necessities for the rest of the population. These ancillary effects of the disease exacerbated the deadliness of these pandemics.

Accounts of this catastrophic encounter often present the Native American groups as passive recipients of a tragic circumstance, without recourse or remedy. While it is true that the pathogens were unknown among Native American populations and had devastating effects on their communities, it was not true that they were passive victims, or that they had no treatment or strategy to mitigate the effects of the diseases, although the cultural logic behind their approach might be nothing like our contemporary approach. Paul Kelton of the University of Kansas has written extensively about the ways in which Native Americans, particularly Cherokee, were able to respond, "retarding mortality rates and curtailing the spread of contagions."[26] Archaeologists still do not know exactly how the various waves of disease happened in all areas of southeast North America.[27] We used to think that the Spanish entradas of the sixteenth century brought devastating waves of pathogens, at least in the areas where De Soto, Luna, and Pardo passed. This interpretation finds little support from historians working on this period, suggesting that it was the sustained contact in the seventeenth century, combined with enslavement, warfare, and other social changes, that triggered the devastating waves of disease. In places like North Carolina and Kentucky, there may have been a much smaller impact until the arrival of European colonizers, who settled in those areas in the late seventeenth century. In any case, eventually, the epidemics and waves of disease arrived.

The devastation wrought by disease after the arrival of the Spanish in the Americas constitutes an unparalleled tragedy in the history of humankind. Within one hundred years of significant contact in the region, the indigenous population had declined by approximately 90 percent. There is very little disagreement in these numbers, and no one seriously thinks the number might be as low as 75 percent or even 85 percent. It was catastrophic in every sense of the word. For Ricardo Agurcia, a Honduran archaeologist I spoke with about the Maya "collapse," the near elimination of the indigenous population constitutes the closest thing to a true apocalypse in the history of the world.

When I teach about this, my students often ask why the transfer of disease operated in one direction, with syphilis being one of the only diseases that traveled from the Americas to Europe.[28] Why were there no analogous outbreaks of diseases among Europeans in the Americas? The answer is complex, but one reason that the European diseases were much more deadly to Native Americans was because Europeans had been living in proximity with domesticated animals in a way that never happened in the Americas. Europeans had cattle, sheep, goats, and pigs, all living around and even in the household, with diseases passing back and forth between animal and human hosts. Native Americans domesticated animals, too, but there was not such a variety of large mammals living in close proximity to people. In the Americas, domesticated animals varied by group and region, and included dogs, ducks, turkeys, or guinea pigs. Large animals were limited to llamas, alpacas, and vicunas in the Andes Mountains of South America. The type of zoonotic disease spread that existed in Europe, jumping from animals to humans and back, did not exist in the Americas at the same scale.[29]

Because people travel, and pass on disease to others, who travel farther, disease moves faster than the original human hosts. Because the diseases traveled faster and farther than the Europeans themselves did, there were few instances where Europeans observed Native American groups living in a manner that resembled societies prior to their arrival. In most cases, waves of disease had already come through and profoundly disrupted life by the time the actual Europeans arrived. This shaped the European view of Native American society, and even influenced our ideas about how many people were living here before the arrival of the Europeans.

While we know that pathogens spread through Native American communities and resulted in great numbers of deaths between the fifteenth and nineteenth centuries, that encompasses nearly four hundred years, thousands of miles, hundreds of societies, and thousands of communities. The details of how this looked in any given area are unique, as are the responses. Communities disbanded, but new ones formed, as did new political units.

In Kentucky, trading patterns and alliances changed with the foreign influence beginning in the sixteenth century. Iroquois raiders came from the north and some degree of conflict probably occurred as the larger political context of eastern North America changed, but the widespread conflict between the various Native American groups reported by the English in the seventeenth century is not supported in the archaeological record. Smallpox probably arrived in the late seventeenth or early eighteenth century.[30] The response to that probably varied from group to group, although we can be confident that 50 to 90 percent of the population died, with the old and young being most vulnerable. During a discussion on this topic, archaeologist Gwynn Henderson pointed out that the young and old represent respectively the future and the collective memory of the group. Some groups could have left the area, while others re-formed or persisted. For this key period, from the late seventeenth to early eighteenth century, there is a complete lack of eyewitness accounts and archaeological data, so we do not know how closely the post-epidemic world resembled the previous one. We do know that in the heavily fractured time in the mid-seventeenth century and beyond, multi-tribal villages were common, as survivors banded together.[31]

In terms of why everything turned out as it did in this catastrophic era, we certainly have the proximate causes: the arrival of Europeans and the spread of disease. However, other factors such as political alliances, trade networks, and movement of neighboring groups had a great effect on the final outcome. This variability across space and time in the effects created by the initial forces demonstrates the difficulty in establishing a clear connection between a cause and effect.

From these examples, the challenges in identifying proximate causes, understanding all the ancillary effects, and finding the threads that link cause and effect become clear. Even when a proximate cause is identifiable, the context in which it occurs can change everything. In my archaeological work, understanding all of the connections and explaining why and how something happened is important. For my purposes here, it is less important. I am not presenting these cases to demonstrate exactly how it all happened in the past. Here, the value of these examples is to compare them with our contemporary fantasies, and the vision of the future that guides our preparation and planning. The common denominator is that these collapses take a long time, although the *realization* that something has been happening for decades may come suddenly. The proximate causes of these crises are intertwined with other events and reactions that occur while things are still in flux. Finally, none of these collapses are complete in the sense that people survive, they rebuild their old groups or build new ones, and that the kind of individual heroics we see in popular contemporary narratives do not save anybody.

CHAPTER 3

HOW THINGS FALL APART

I had a fever and all the symptoms of a bad cold. I was in Honduras, just returned from presenting my research at the University of Costa Rica. I felt bad the entire trip, but it seemed to go away, and on the day I flew home to Kentucky, I felt good most of the day. The next day, I developed a very high fever, around 104 degrees. I thought I might have developed pneumonia, and I went to the hospital. It turns out that I had malaria, which I thought would be easily recognizable from the cyclical fevers. It was not that easy to identify, however, and it was not at all clear to me that I had malaria until my diagnosis, a couple of weeks after symptoms had appeared. I was admitted to the hospital in Lexington overnight while they made sure that there was not something else wrong too. Over the course of that night and the next day, a dozen interns, nurses, and doctors came in to look at my chart and ask questions. It seemed clear that everybody in the university hospital had heard about the case of malaria, which had not been a common illness in Kentucky since the early twentieth century, when thousands of cases per year were reported.[1] So unique was this, in fact, that several months later as I was back in the hospital with my wife for the birth of our youngest child, one of the attending physicians recognized me as "that guy with malaria." Each of the people who came in to "check my chart" ended up asking questions, and many of them boiled down to a desire to know what it felt like when the tremors started (associated with the release of toxins into the bloodstream by the protozoa that cause the disease). I could not explain it. More than that, however, I could not explain to my family or friends over the next few weeks the mental effects of such a debilitating illness. I lost interest in everything and vacillated between depression and apathy. That was by far the worst part of the experience. I could explain the cause of my condition, the *Plasmodium vivax* protozoa. I could not communicate what it was like to live through the illness caused by that parasite.

I can imagine something that I have not experienced, but I do not really know what it would be like to live through. I remember describing the feeling of a broken bone to somebody who had never had one. I could not convey exactly how I could tell where the crack in the bone was, nor how it felt. I could describe the cold, sharp pain that was slightly sickening, but I could tell that those adjectives barely even made sense if you had not experienced it. Behind their eyes, I could see the question: *What is a cold pain?*

Finding a cause for a catastrophic event is just the start. Ultimately, we are interested in what these events looked like for the people who lived through them . . . or tried. For archaeologists, it is difficult to comment on individuals. We deal with patterns, with groups, and our ability to focus on an individual is often limited. Extrapolating from what we know about the larger collective situation, we can try to tease out the realities of living through a radically transformative period.

The Classic Maya collapse was not a homogenous, uniform event. Different parts of the region were treated or were affected differently. We also see that it was a long process, longer than we often imagine. It appears that the collapse happened over a century or more, although more rapidly in some areas. It is clear that a variety of causes for these events existed, including environmental causes related to deforestation or drought, but in the end, there was a whole slew of things that went wrong. This culminated in a loss of confidence in the systems that were in place. Even if there were a singular proximate cause that set it all off, it might capture the nature of the event better to think about it as a multicausal event. Not only was it multicausal, but the collapse may also have looked very different over space and time. There may not have been a singular experience to untangle.

As important as this may be for thinking about future catastrophes, the fact is that the Classic Maya "collapse" was not complete, at least in terms of population. Communities continued to exist, albeit as smaller versions of those that had populated the large cities. Certain activities ceased, including monument production, and people moved out of the cities. Population declined in the area, but this took place over the course of a century. Moreover, millions of Maya still live in the region. I always return to the contrast between our fantasies of going it alone with the reality that communities—groups beyond our families—will always be how we survive.

So, how did life change at the end of the Classic Period for the people living in the area? First, there is strong evidence of a population decline in the southern part of the Maya area, and only a handful of sites were able to celebrate the new *baktun*. Patricia McAnany comments on these changes, saying, "So we know that the Maya 'divine rulership,' that system of governance, got into trouble. People lost confidence in it, maybe much like people losing confidence in our government today. So, at that point in time, people began leaving. We see the end of that form of rulership."

While divine rulership ended, and people abandoned cities to the south, there is evidence that the northern part of the region followed a very different trajectory. McAnany continues: "I think what's very often overlooked is the fact that the northern lowlands—although they, too, suffered political privations at the end of the Classic Period—experienced a resurgence in the Postclassic, and that is where a lot of population was located when Spaniards came onto the scene in the sixteenth century." The centers of power and population shifted. For your average person in the southern part of the region, the political system that had been in place for centuries began to unravel, and the locus of this power, the cities, suffered. The nearby urban areas would have declined and ceased to be as important as before. Families and smaller collectives ended up as rural communities. In terms of Mayan spiritual life or worldview, we know that continuity exists between the ancient and the contemporary. Belief systems are always challenged, but the loss of faith in a particular divine rulership did not undermine the core religious beliefs.

McAnany summarizes the events of the ninth century in this way: "Whether [divine rulership] got into trouble because of too much warfare or too many droughts—frankly, the evidence is not clear on that—or whether it was something else altogether, it did get into trouble. But it was a system that was put in place sometime in the early Classic [250 CE or so]; some would argue even earlier. It was a system of governance that had lasted somewhere between five hundred and one thousand years, depending on what measures you want to use for the onset of divine rulership. And I think if our democracy is around five hundred to one thousand years from when it was founded, we will definitely consider it a success."

One of the focal points for both Maya scholars and residents of the area is the resilience displayed by the Maya in the face of these changes. Religious traditions, ceremonies, and the use of traditional language continue to this day. We tend to think of the loss of divine rulership, the decline of the large cities, and the cessation of monument production as some sort of great loss, but we might consider that the changes may have freed the population from a cumbersome system. A culture that creates spectacular artifacts and spectacular monuments may not create a spectacular quality of life for its people. The measure of success might not be found in the continued existence of a particular political or social system, but in the lived experience of the common folks.

It helps to look at the Maya collapse from other angles. Work in adjacent regions helps reveal the larger context.[2] What happened when the Classic Maya civilization transformed looks different as you move around the region. It even looks different as you move around the homeland, the area where they lived. It looked very different in areas just outside of the Maya homeland, such as where I worked in eastern Honduras. This region, only 200 kilometers or so east of the Maya homeland, is fundamentally different in some ways, both now and in the past, but very similar in others. The collapse looks very different there.

Currently, the east is much less densely populated than the west, includes huge tracts of rainforest, and is home to indigenous peoples who can trace their language and ancestry to lower Central America, Panama or Colombia, rather than to the north, to the Mesoamerican tradition that the Maya shared. These differences existed even at the time of the collapse. We know this by looking at ways they decorate pottery, build structures, and lay out their cities. There are a few similarities, but anybody looking at the things left behind would see the difference immediately.

I first traveled to Honduras in April of 1991. April is the driest, hottest month, and people take advantage of that and burn agricultural fields. Everything burns, it seems. After a week or so of getting paperwork in order, I left the capital city of Tegucigalpa with George Hasemann, my mentor, friend, and the archaeologist who invited me to work in Honduras. We drove east, through the brown countryside, obscured by smoke from the fires set to burn off the fields, clearing overgrowth and fertilizing the soil in a modern version of the millennia old slash-and-burn agriculture. George drove us in an old Land Cruiser from the Honduran Institute of Anthropology and History, with all the gear I would need for the next three months in the back. Darkness fell as we reached the end of the paved highway, and we continued into the darkness on a dusty, rocky road. Only the roadway was dark, though. The hillsides were on fire. Long snakes of fire, the edge of advancing flames, stretched along the roadway and up the hills. We drove through flaming doorway after doorway. Over the diesel engine, the fire sounded deceptively alive. In the aftermath, however, daylight revealed that nothing remained alive, and one blackened field after another led to the edge of the pine forest, then under the pines, coming to an end at the hardwood rainforest. We stopped in a town at the intersection of agricultural land, pine forest, and rainforest. The town was dark, as electrification was still several years away. We slept in the only lodging in town—a tin shack divided into tiny rooms with pressboard walls. We woke a man who was sleeping in the front room, and he gave us each a candle and a box of matches, then led us to our rooms. I stood beside the small bed and unrolled the thin mattress and laid it on the springs. Picks, shovel, screens, a stadia rod, tripod, and gear cases took up the narrow space beside the single bed.

George stayed with me for two days, helping me find local guides and showing me the protocol for recording archaeological sites. The first day we talked to local community leaders. The police chief for the area asked if I had a firearm. I assured him that I did not. That answer worried him. The next day we went to see a large archaeological site that most everybody in the area knew about. This site was located in a pine forest, and consisted of many earthen mounds, some low and others rising to an impressive five or six meters in height. I stood there trying to understand what I was seeing, and thinking about how to map all of it, when George came back from looking around. "That is a ball court," he said, excitedly pointing, and I struggled to see it.

This was significant. Ball courts, in the Maya world and throughout Mesoamerica, are locations for games played with rubber balls, but are also much more. They had enormous ritual utility and symbolic content, more like a cathedral than a basketball arena. We were firmly outside of Mesoamerica, and outside of the places in which you would expect to find these structures. Explaining why people had built ball courts in this part of Honduras became a focus of my research. How could something so symbolically loaded be transferred from one region to another? How can you transplant something central to one worldview into another? It was not just a matter of architecture: it reflected a vision of the cosmos, central to the religious thinking of the people who built them.

Over the next several years, I conducted archaeological surveys all over eastern Honduras, recording a couple hundred sites, almost all of which were known to local residents; I did not "discover" anything, in that sense. We have a saying in archaeology: "It is not what you find; it is what you find out." We found out quite a bit.[3] With the help of the Pech with whom I worked, I was able to figure out a timeline for the developments in that region. Given the ball courts we found, the relationship between eastern Honduras and the Maya world became particularly interesting.

Around the time that the Maya were declining in western Honduras, Guatemala, and elsewhere, eastern Honduras ascended. People built more and larger sites, suggesting a population increase and a general increase in power for the elites who were running things. Whether this means that life was better or worse for most people living in eastern Honduras around 1,100 years ago, I cannot say. I can imagine it both ways: living in a group that has gained power could be advantageous, but living under elites who'd recently grown more powerful could be oppressive. It seems to be true, though, that the decline of the powerful polities in the Maya area coincided with the increase in size and complexity of groups in eastern Honduras, and I posit that the power vacuum left by changes in the Maya area facilitated the increased population and prominence of neighboring groups. So, from the perspective of a resident of eastern Honduras at the time, the decline of the Classic Maya civilization would not have been an apocalyptic event at all.

Eventually, in eastern Honduras we also see declines. One of the mysteries to me is what happened just before the Europeans arrived. In most places I researched in eastern Honduras, we saw a steep drop-off in the building of large sites after 1300 CE—something like what we saw in the Maya area around 900 CE. Looking at the archaeology of the region, I see very little evidence of what was happening in the last few centuries before the arrival of the European colonizers. The documents from the Spanish in the sixteenth century, however, suggest that the area included relatively large settlements. While colonial-era documents demand scrutiny and must be interpreted and understood in light of the motivations of the authors, so many documents hint at the same thing that I feel certain that there was a sizable population in the region when Europeans first arrived. The settlements from this period are not readily identifiable to me as an archaeologist, however. There are many possible explanations. Perhaps the stylistic clues from the artifacts that I use to assign a time period to a particular site are insufficient, or insufficiently fine-tuned, to distinguish 1500 CE from 1200 CE. Perhaps sites were occupied over a long time, and earlier occupations might obscure the later ones. Future research will help answer these questions.

When I look at the larger context for the Classic Maya "collapse," the futility of looking for a single cause becomes obvious. Looking at neighboring groups brings into focus the fact that the "collapse" affects a particular set of systems—one groups fails, and a nearby group prospers. A multitude of causes put pressure on a political, religious, or economic system, and the particulars of that situation and the response to the pressure determines the outcome. The reaction to a particular stimulus can vary enormously from place to place, or time to time.

Sorting out proximal causes for the decline of the Western Roman Empire is as difficult to do as for the Classic Maya "collapse," and the addition of textual data to the archaeological data does not make it any easier. Ultimately, though, identifying a proximal cause is not that critical to understanding what happened once it was all set into motion. What this decline would have looked like to the people living in the declining empire is also complex. The archaeologists I spoke with and the literature agree on the gradual nature of the decline. It was "a progressive transformation, a fading," in the words of archaeologist Riccardo Montalbano. It was a slow process, not abrupt.

The gradual, almost stealthy nature of the decline has been one of the defining features of this historical period. The historian Arnaldo Momigliano wrote about a "*caduta senza rumore*," a "noiseless fall."[4] Riccardo Montalbano continued with this thought: "That is because the contemporaries probably didn't experience it as a trauma. To give an example, even after 476 CE [the date often cited as the year the empire fell], the Roman jurisdiction continued to exist, as did many traditional institutions and the Senate." With such a slow decline, and the persistence of so many elements of the disintegrating empire, the notion that a collapse was under way might have been completely foreign to the people living through it. This phenomenon, not recognizing or even denying a collapse, seems to be common throughout human history, and many contemporary observers wonder if history is repeating itself in this regard today.

Almost all archaeologists working in the pan-Mediterranean region reject the term "collapse" for this scenario. Even with a more nuanced concept, we need to take care with our categorization of what happened. As in all of our discussions of the past, or the future for that matter, the vocabulary makes a difference. Montalbano talked about our tendency to create rigid categories that restrict our understanding of the past. He notes, "When we talk about transitions between different historical periods, the risk is always the inappropriate use of some stiff categories such as *discontinuity* or *continuity*, the latter generally applied to phenomena that are simply *persistences*. It is clear that, even after the 'official' end of the empire (476 CE), many constituent elements of Roman culture survived the fall of the empire: For example, Roman law, which is the basis of modern European law."

While the decline and fall of the Western Roman Empire was a significant event, obviously, it might not be the most significant in the region. Even greater changes occurred at other times in the history of the area. The spread of Islamic civilization, for instance, had a more significant effect on the region than the decline of the Western Roman Empire. Montalbano notes, "In my opinion, the most profound and radical change since the fall of the Roman Empire is the development and spread of Islamic civilization, starting from the middle of the eighth century CE. Until the seventh century CE, all the shores of the Mediterranean shared—albeit with differences—the same fundamental traits (religion, habits, etc.). We can still talk about a Mediterranean unit. It is only the sudden burst of Islam that breaks this unity, by opening a new chapter in the history of the Mediterranean basin."

That sort of pan-Mediterranean viewpoint is critical to contextualize the decline of the Roman Empire. In the same way I look at the Maya from outside their homeland, we can look at the decline and fall of the Western Roman Empire from a similarly broad vantage. I spoke with Dr. Kara Cooney of UCLA about these events, seen from the perspective of somebody who works primarily as an Egyptologist, and has looked at this decline and fall from a perspective at the edge of the empire.

Given the long, slow nature of the decline of the Western Roman Empire, I asked Cooney if she had identified a moment when events crossed a threshold and the decline became a collapse. "As an Egyptologist, as a social historian, I tend to avoid dates and I tend to look at these events through the *longue durée* [the long term] anyway, and the *longue durée* of the fall (of Rome) is the most interesting and essential part of it. Picking a date is part of what ruins it. By not artificially picking a date for the moment of collapse, we can better understand how long this took, and what that would mean for the people going through the transition." We have to look back much further, often, to see the origins of an event, and it is in an appreciation for that long, drawn-out process that we can begin to understand what people living through it might have experienced.

How would this collapse have affected the average person in the Western Roman Empire? There is no singular answer to this, since experiences depended on socioeconomic status, ethnicity, geographical location within the empire, gender, and other particulars of the person under discussion. The Western Roman Empire spread throughout the western half of the Mediterranean, from Albania around to Spain and back east, nearly to Egypt. The decline might have played out very differently on the outskirts of the empire. The degree to which somebody living on the edge of the Empire would have had a significant, daily articulation with Rome would be unlike that of a resident of the seat of the empire. The philosopher Synesius of Cyrene, writing in 408 CE, talks about citizens at the distant edge of the empire, in Libya. He speculates, "No doubt men know well that there is always an emperor living, for we are reminded of this every year by those who collect taxes; but who he is, is not very clear. There are people amongst us who suppose that Agamemnon, the son of Atreus, is still king."[5] In fact, Flavius Honorius was the emperor of the Western Roman Empire at the time, and Synesius's comment is suggestive of the disconnection with the empire except for utilitarian interactions, such as taxation.

For many people the day-to-day realities of existence might have been more important than the changes happening at some other level. Cooney puts it this way: "In some ways, when you're talking about the bottom of society, it affects everything and nothing. Because the bottom of society has to keep on keeping on. They have to farm. They have to produce. For the bottom of society in an industrialized Rome, in an urban context where the city itself might be uninhabitable in terms of water or sewage, those people would have to go on the move. They would have to find other patrons, find other things. They have no other means to feed themselves."

What is negative for one group might be positive for another, and somebody living in proximity to other groups could have options during a collapse that somebody located in the center of the empire might not. Assigning positive or negative value to any of these developments might be the wrong approach. The collapse of a system might create great suffering on one hand, but also allow opportunities for those previously denied them. The rebuilding process is also complex and results in new opportunities for some, new strategies, and even new technologies.

As we saw with the Classic Maya, we see that the decline and fall of the Western Roman Empire was complex, multicausal, spread out over a long time, and nowhere near the complete collapse so characteristic of our popular narratives. The people are still there, as are many parts of the political and social system. This change did not create a centuries-long "Dark Ages" in its aftermath; in fact, the subsequent, postcollapse period from the sixth to tenth centuries CE was not particularly "dark" for the region. Focusing only on what is lost distorts our perception of the event.

The last of the three examples I am exploring here, the events in eastern North America in the fifteenth through eighteenth centuries, has much more documentation than the others. In places where the Europeans arrived in the fifteenth and sixteenth century, we have accounts of the encounter and the ways in which things progressed. In other instances, where disease spread faster than the colonizers, no documentation exists and archaeological data becomes the primary way in which we can understand the scope of the changes that took place in the first centuries after the arrival of Europeans in the Americas. This is a true apocalypse for much of the population. Such disruption and such a startling number of deaths happened for which there is no parallel. Perhaps the closest thing to a similarly devastating type of apocalypse would be the enslavement of 12 million people in the Americas, beginning almost immediately after the arrival of the Europeans and continuing for more than three hundred years. We know that the devastation wrought by European diseases fundamentally changed Native American societies. Such widespread disruption left virtually nothing unchanged. Even in less devastating events, such as the Covid-19 pandemic, we see the ways in which our interconnectedness makes everything susceptible to change.

While life changed, people persevered, and the resilience of people in their traditions is astounding. Native Americans survived, and while they certainly changed, they maintained traditions, created new traditions, and re-formed themselves into meaningful groups. Just like the Maya, indigenous people in eastern North America never truly disappeared; their cultures persisted. We can also see that even in this kind of near total collapse, which is the closest we get to some of the kinds of fictional apocalyptic narratives that we invent, it still did not look like what we imagine in our popular narratives. While people dispersed, they continued as groups in communities. They continued to farm. They continued to trade. They continued to engage with the complicated, messy, changing world that they were living in even in the midst of disaster. The simple, decluttered life of our fantasies, where we get to be alone with our chosen group, was not the reality, even in a situation with massive demographic collapse of more than 90 percent over the course of one hundred years. This depopulation was so great that it killed 10 percent of humans on the planet and cooled the Earth.[6]

By the mid-eighteenth century, the Native American population had declined to around 6 million, meaning that the population had originally been more than 60 million. Populations decline for many reasons, including lower fertility rates. In this case, however, most people died during waves of disease that came through an area. What happened to the indigenous populations of the Americas surely qualifies as an apocalyptic event, and the proximate cause of the catastrophe was disease. How this played out historically differed radically from most of the outbreak narratives we reviewed earlier.[7]

We are still in the midst of the Covid-19 pandemic as I write this. The death toll is high, and the impact is profound, and all with a death rate exponentially lower than that experienced by Native Americans in the sixteenth through eighteenth centuries. Even with our current experience, I can scarcely imagine something as dramatic as that. I teach at a small, residential college in Kentucky. Typically, most students live on campus. I asked my students what would happen if one student in the dorms died of an unknown disease. They hesitated, and I asked, "What if it was two? Three?" At that point, there was no hesitation. They would be gone, either on their own accord, or at the insistence of parents. Everybody who was able to do so would leave campus. Four students? Five? Everything shuts down.

Even relatively limited epidemics can have major consequences. Where I live in Lexington, there is the legend of William "King" Solomon. King was a drunk, arrested so many times for public intoxication that the city sold him into servitude for a period of nine months. During this time, in 1833, a cholera epidemic hit the city. Within two months, approximately five hundred people died, out of a population of 6,000. Stores shuttered, and people fled the city. The gravediggers fled as well. With so many dead to bury, King Solomon took up the call. Legend has it that his propensity for whiskey and almost total abstinence from water protected him from the waterborne pathogens. Upon his death, townspeople honored him with a statue in the city cemetery. King Solomon's life changed, from outcast to hero, as did the lives of everyone in the region. In this case, approximately 8 percent of the population died, or 1 of every 12 people. Imagine what it must have looked like for Native Americans when 11 of 12 people died.

During the lockdowns of the Covid-19 pandemic, a relatively low death rate prompted dramatic changes. The death rates were much higher than any seasonal virus we had experienced (at least ten times higher than the average flu, according to best estimates), but nothing like the waves of disease that swept through Native American communities in the sixteenth and seventeenth centuries. The near annihilation of Native American cultures during that time period provides a glimpse into societal responses to massive demographic collapse. Waves of disease passed through the populations, sometimes killing 75 percent of a village in a single winter. Typically, a given area lost around 90 percent of its population within one hundred years of the arrival of Europeans and their livestock.

Like the other archaeological case studies we looked at, the concept of resilience comes up as often as the concept of collapse. In 1847, in the middle of the potato famine, the Choctaw Nation gave a gift of $170 to help the people of Ireland (the equivalent of around $5,550 in 2021 dollars, although I imagine that figure does not capture the generosity and sacrifice represented by the gift). The Choctaw had been displaced during the Trail of Tears, like other southeastern US groups, only fourteen years before, where around a quarter of the displaced people died. That is resilience.

Looking at the response by Native Americans to the truly apocalyptic catastrophes that befell them, from disease to genocide, the resilience of these groups stands out. They changed and adapted in ways that were not merely reactive. We can see how they retained concepts of community and humanity, and even strengthened their commitment to those ideals in the face of horrific outbreaks, barbaric treatment, and marginalization that continues to this day. Once again, our rugged individualistic fantasies of going it alone after an apocalypse do not resemble what we see historically or archaeologically. Communities rebuilt, and while short-term situations surely ran the gamut of horrors, people coalesced into groups again and persevered collectively.

For Native American groups in eastern North America, the experience of the numerous catastrophes that occurred would have varied greatly from place to place, but some salient similarities can be identified. First, even in the cases where so much of the population was devastated by disease, people persisted as members of a community. Outside of isolated or short-term situations, people retained, re-created, or sought out new communities. Also, the power of these communities to react to what happened never faltered. Strategies for dealing with disease emerged, and some of these had demonstrable, positive effects. The ability to shape the present and the future, the agency of the people, survived the waves of disease, slavery, and genocide.

The past does not look like our fantasies. The systems people orchestrate for themselves, or impose on others, create, allow, or exacerbate crises. Ultimately, these complex systems fail or are replaced. The aphorism that there are no natural disasters, just natural hazards (as overused as it is) captures that.* Our reaction and the systems we put in place beforehand shape how it all plays out.

---

* The expression might be overused, but it is true and it is useful to remind ourselves of the role we play in how a natural hazard affects people. There is a campaign dedicated to spreading this message at www.nonaturaldisasters.com.

In thinking about the next apocalypse and looking to the past for guidance on how it might unfold, I keep returning to the vocabulary we use to describe the events. In particular, our concept of collapse or fall, or even apocalypse, seems to shape our imagination and set the parameters for the discourse in which we engage on the topic of the future. Could our ideas about future apocalyptic events be shaped by the repeated use of terms such as "collapse" that imply to us a failure of community, and the need to take care of ourselves on our own? Revisiting the concept of collapse helps answer that question.

Scholars understand the complexity of these situations, and a dissatisfaction with the concept of collapse when applied to past civilizations is not new. Sociologist Schule Eisenstadt argued that collapse rarely occurs, if we define it as a complete dissolution of an existing political and social system.[8] Archaeologist Joseph Tainter wrote extensively on this subject and concluded that collapses did not happen.[9] Things changed, of course, and there were catastrophes and crises, but not true collapses of the kind we imagine. More recently, Patricia McAnany and Norman Yoffee edited a book, with many archaeologists discussing supposed collapses and coming to a similar conclusion.[10] In fact, the focus shifts from what fell apart to how people restructured themselves.

Some archaeologists, however, are not so keen to throw out the idea of collapse. Takeshi Inomata discussed this with me and noted that he sees evidence of a true collapse in the data about the Classic Maya, with demographic declines of as much as 90 percent. He suggests that we think we are seeing a gradual decline because we do not have good, tight timelines for these events, which he thinks happened more rapidly. Guy Middleton also supports the use of the term, suggesting that other terms such as "decline" or "transition" "may mask the real horrors experienced by people during particular historical events and changes, even as it successfully describes changes in political, social, and material culture."[11]

Archaeologists often prefer to talk about the resilience of groups that have undergone dramatic change rather than the collapse of those groups. We focus on what remained rather than what changed. We explore the continuity in addition to the discontinuity. We could look at certain groups, like the Nazis in Germany in the 1940s, and see that the collapse of a particular political system does not necessarily constitute a tragedy. A collapse also might not constitute a failure. A political system that endures for a millennium can hardly be called a failure, regardless of how it ends.

One of the issues raised here involves our ability to identify the very beginnings of a significant change. Sometimes, the progression of a collapse happens so slowly that we fail to appreciate how far in the past the origins of a particular change might be. In my conversations with Kara Cooney, this idea came up several times. We talked about the difficulty in identifying the moment when something begins. I related my conviction that much of what we see in the rise of neoliberalism today started with the election of Reagan. She agreed, but suggested that we could look further back, to Nixon or to Jim Crow, or to the creation of the slave state. We could even go back to the beginnings of significant inequality with the shift from foraging to agriculture, some twelve thousand years ago.

Not only can the process of collapse be very slow, but there can be different stages or steps to the process in which people's lives are ever-changing. Cooney notes,

> [The term "collapse"] is a quick cipher, and it's helpful. Everyone knows what it is you are talking about in the large scheme of things, but the word can obfuscate in some ways. Each collapse is different. It depends on what you're talking about collapsing, because, while certain systems will collapse and certain patrons [ruling elites for which people work] will fail, the patronage

> system generally stays in place, just in a decentralized fashion with people moving from one patron to another, where people are on the move literally and trying to find new patrons and new systems.
>
> So it's like a divorce . . . where everything gets thrown up in the air, all of the quotidian details of what you will eat, where you will live, your paycheck, your child, every little detail becomes up for grabs. And it takes a long time, as you throw it up in the air, for it to settle down into something new. So, you're breaking up with the old partners. Then you will get back together with new ones. And it takes a couple of years.

She notes that the concept of collapse might apply best at certain points in the process of transformation. She continues:

> I think that the word "collapse" best applies to the beginnings of the breakup, when everything's fallen apart and people are all on the move and nobody knows what to do. And it describes the unknown state of confusion, the fog of the collapse, before the reconstitution, which starts to happen pretty soon after. People start to put some pieces back together and very nimbly re-create decentralized, messy systems to replace what they have lost. But those moments of expecting one patron to give you what you need and not getting it, that's the part that people I think apply the word "collapse" to: Where is my Social Security check? It's not even arrived. For us, we would be asking, How do I eat, what do I do, I can't pay for my house. I'm on the move. A Dust Bowl sort of collapse. And yes, there were ways of putting things back together, and people did find new patrons, but it's that initial part with the fear and the panic that I think really evokes what you're getting at.

A collapse might be several collapses, in fact. It drags on, and things fall apart and are reconstituted only to fall apart again, perhaps in a different way. Cooney sees this as being true for Rome. "For the fall of Rome, I think one of the reasons it's so difficult to pinpoint is that you get the collapse of a certain kind of Roman state and then a rebuilding into something else, with different kinds of patronage. And then that collapses, and then a rebuilding again, and then that collapses and the rebuilding again. And it could arguably happen ten to twelve times and in quicker succession, as things evolve, as things move. You get this series of collapses. That's the way I prefer to look at it."

So how do past catastrophes compare with the fantasies? Stark and significant differences exist between the situations we see in the past and some of the apocalyptic images we entertain. These disparities suggest that the next apocalypse will not happen as we think it will. In fact, if it is a gradual decline, as we saw in Rome, we might not even recognize it.

Most of all, history suggests we are preparing in the wrong way. We should be thinking communally, not individually. In not one case, even in the face of great demographic collapse, did people stay living on their own in tiny groups for any amount of time. We like to focus on individual survival, but we need to think about restoring structures and systems. By that, I am not suggesting we re-create them as they are, as that may be neither desirable nor possible. In the narratives of the past, we have no heroes. The metaphorical battles are won or lost by a community, not as individuals. We can also see that for every leader, there are many followers. Most of us, in most contexts, need to be part of a community, working together. The followers, not the leaders, will rebuild everything.

Some things remained constant, however. Cooney talks about the social order we see in the past: "What never happened, and has not happened until recently, is that the collapse ushers in an anti-patriarchy. If all of complex civilization from the advent of farming and herding is patriarchal, and most archaeologists would agree with that, then these collapses actually enhance those patriarchies in their most basic way." During these crises, she sees the patriarchy strengthening, and paternal masculinity expanding. "Strong man protecting his wife and daughter, and that warlord sensibility. That patriarchal male dominance, binary male/female roles, those things get more intensified, sharply delineated in a collapse rather than in those degenerate pre-collapse periods."

The reconstituted patriarchy then looks for scapegoats. She continues: "Looking back at what it had left [those representatives of the patriarchy] remember [the system that] allowed people to be nonbinary in their sexuality, that allowed women into some positions of professional power. They look back at that and they see it as the reason for the collapse and blame it on the people themselves. So, the collapse can be horrific for anybody looking for a liberality of mind that goes beyond a patriarchal expectation of survival."

In the next sections, I move from this critique of apocalyptic "collapse" narratives as applied to the past and look at our tendency to do the same thing with the future. Examining the past, in my three main examples and in dozens of others I have not mentioned, I am alerted to the potential misapprehensions when it comes to future apocalypses. People and cultures persist with much greater continuity and, necessarily, interrelatedness, than our overemphasis on collapse conveys. If the past is prologue, continuity will remain the feature of humanity, even in proximity to catastrophe. The futures we imagine do not reflect this continuity, however. Our tendency to highlight discontinuity while obscuring continuity has encouraged misreadings of the future.

PART II

The Present

Envisioning the Apocalypse

We love to think about the end of the world, as we know it. We imagine it, and we read and watch other people's visions of it. We immerse ourselves in these fantasies. It is cathartic, maybe even enjoyable. Preparing for the apocalypse has become a hobby or a lifestyle, not a grim task we resign ourselves to undertake. We even watch and read about other people preparing, or "prepping," for the next apocalypse. We might be concerned about a future catastrophe, but we sure like to imagine it. Like wiggling a loose tooth, it hurts so good.

Sometimes pain just hurts, however. Our worries about the future of humanity and the planet go beyond the normal concerns that most people have about their personal future. We worry that the future holds tremendously catastrophic surprises, or maybe inevitabilities. I can currently feel the fear and dread that world-changing events will overtake us, as they have at times in the recent past. During the Cold War, we feared nuclear destruction. Now it is climate change, or the rise of authoritarianism, or a pandemic. Sometimes, all of the above.

Despite the real fear, we continue to fantasize and create stories about civilization collapsing, an interesting and revealing paradox. On the one hand, we get pleasure and catharsis from thinking about life during and after potential cataclysms. On the other, we dread such a scenario, and our vision of the future leaves us scared, depressed, or hopeless.

I have seen both sides of this bipolar approach to future disasters during the wilderness survival courses I teach. I began teaching them shortly after hearing about the death of a man who took a wrong turn and became lost and snowbound with his family in a remote part of Oregon. After waiting in their car for about a week, the father set off on foot to find help and died. Had he made a few different decisions, he might have lived through the ordeal. I thought about my students, driving home for Thanksgiving or Christmas break, and I began to give presentations about basic steps to take in a survival situation.

My training in these matters came from my years living and working in remote parts of Honduras. My research took me deep into the rainforest, and a small group of us would travel for a few weeks on foot before returning home. Everybody I lived and worked with, members of the Pech indigenous group, had grown up there and they knew the forest well. We documented archaeological sites that they knew (and they knew them all, it seemed), as well as a few sites that were marked on old topographic maps. Sometimes we had mules to help carry the load, or we would travel in a dugout canoe along the rivers in the rainforest. These trips lasted weeks, although on the longer treks we would return to some small town to resupply. Out of necessity, we improvised a lot in the rainforest as well as back in the village where we lived. We made rope from tree bark to lash together a balsa-wood raft, we built roofs from palm fronds, and we hunted and fished to supplement the food we carried. I could not do any of these things at first, of course, and my efforts were always amusing to everybody else. Some things I could do well, with repetition, including difficult tasks like making fires in the pouring rain. After a while, these became second nature.

During our trips to the rainforest, we slept under tarps or in hammocks and ate beans, rice, and spaghetti. We carried flour to make tortillas. Despite the name of the region, the Mosquito Coast, insects were typically not too bad, certainly not bad when compared to most summer nights in Kentucky. Sometimes we would run into particularly bad areas infested with biting flies and gnats, and I started using a hammock with a built-in mosquito net.

The principal difference between traveling through the forest and living in the village was the amount of walking and the heavy packs we carried. Other than that, everyday tasks in one of these camps were not that different from life in the village. There was no electricity or water system in the village, so we did many of the same things on our trips away. My wife had grown up in Honduras in a similar village, and the things I now teach as "survival" or "wilderness" skills amuse her. She makes fun of my survival courses. "It's like teaching a class on unloading the dishwasher and turning on the oven," she jokes. Well, "jokes" is not the right word, actually. She means it.

As a child, I grew up in Lexington, Kentucky, and later Knoxville, Tennessee. My extended family lived in eastern Kentucky and West Virginia, in the Appalachian Mountains. I spent a lot of time out in the woods—or "up in the hills" as we'd say. Later, I worked conducting archaeological surveys all over eastern Kentucky, and some in West Virginia. That involved a lot of walking through the woods and looking for archaeological sites. I often spent all day outside in the forest, both in my childhood and later in my professional life as an archaeologist. This does not automatically result in having any particularly useful wilderness survival skills, however, apart from an ability to move through that kind of landscape and being familiar with it. I was inspired to learn about wilderness skills after participating in a documentary about survival in the rainforest.

The BBC invited me to participate in a documentary titled *Trips Money Can't Buy* featuring Ewan McGregor, in which my friend Jorge Salaverri and I designed a trek for McGregor and bushcraft expert Ray Mears through the Mosquito Coast of Honduras. During the course of filming, which centered on Ray demonstrating survival techniques, I realized that I had already learned and practiced many of those skills during my time working and living in remote areas, even though I had not thought of them in that way. Ray is an active student of the people and places he goes, and he demonstrated how becoming proficient in surviving in a particular environment entailed much more than merely being there for a while. He studied the locals' techniques for dealing with common situations, like making rope, getting clean water, or making fire. He studied the types of resources that were available in a particular environment, such as fungal-fighting tree bark or a plant that would purify water through capillary action. Every evening, Ray talked extensively with all of the local folks in our crew about topics ranging from clothing to weather to agriculture.

I began teaching some of those skills to my students, eventually offering courses to the community. It was during these classes, years later, that I saw the two sides of our typical responses to a potential catastrophe. It became apparent that envisioning an apocalyptic future, where it all falls apart and has to reset, can be dreadful and liberating at the same time. A college student in one of my survival courses raised her arms in mock invocation, inviting the universe to *bring it on*. The first casualty in her fantasy would be her student loans. For her, that comment represented something greater than one small positive in an otherwise negative situation. For her, the scenario was not so much societal collapse as a societal reset. For all the negatives, the change also meant freedom and liberation. Her imagined apocalyptic future provided an escape from debt, from environmental degradation, from a tired relationship, or from the ennui that accompanies modern life. We imagine that the apocalypse might serve to change a fundamentally flawed society. In some religious contexts, this notion of the apocalypse as a corrective to a decadent society, as divine retribution, is central to their narratives.

Worry accompanies this enthusiasm for change. Among my students, for instance, I hear trepidation and uncertainty more than excitement when we discuss the future. While some voice their support for a radical change as a way out, most of them dread the possibility of a truly apocalyptic event. Some worry about personal safety. Others have children and are concerned about keeping them safe. The requests for children's survival courses and the questions the parents ask makes it clear where their concern lies. They ask, "How long can a young child go without food? Are smaller people more prone to hypothermia?" I hear the question behind the question: *Will my children be safe?* I know they are afraid.

This fear goes beyond concerns about safety. Some fear the possibility of a future stripped of beauty, art, or play; they fear that life might become utilitarian, that relationships might become transactional, and that much of what makes life worth living could disappear. Some are concerned that a collapse of the rule of law invites a survival-of-the-fittest model, with bullying on a grand scale.

On the other side of the coin, people who prepare for disaster—aka "preppers"—spend time and money in anticipation of some imaginary apocalypse and believe that an apocalypse would vindicate their strange obsession and validate them as their archaic skills become important. They are not just preparing for a disaster. They are preparing for a radically changed society where they will be valued.

This coexistence of a desire for radical change, even disastrous radical change, with tangible fear of the future reveals a lot about how and why we envision certain scenarios. Our penchant for imagining an apocalyptic future is more complex than merely wanting change but fearing it at the same time. That seems too tidy, and it does not capture the particulars of this situation. It is not just any apocalyptic future that people fantasize about and fear, but a future that fundamentally transforms our world. As with the past, picturing discontinuity in the future is seductive. Part of this focus on a future that portends discontinuity, in which our contemporary reality undergoes significant changes, relates to our real and increasing concerns about climate change, political shifts, or warfare. The multitude of narratives about the apocalypse that we see reflect this predilection toward a dramatic future collapse. Zombie movies, dystopian novels, eschatological religious thinking, and the litany of prepper-related media embody our fantasies and fears. We increasingly imagine a future where everything changes. Sometimes, it's a sudden and complete collapse. Other times, a dystopian future sneaks up on an inattentive public. We fret and fantasize, and some even plan for it. In this section, I survey the ways in which we imagine a future apocalypse. I look at how we prepare for the worst, at the world of survivalists and preppers. I examine apocalyptic narratives in print, film, and from religious traditions, where we see how contemporary concerns shape those ideas and affect how we think about the future.

CHAPTER 4

APOCALYPTIC FANTASIES

Water was up to my knees, which was disconcerting because I was sitting in the passenger seat of a truck. I could feel the wheels trying to get a grip on the river bottom as my friend Leon tried to drive across, but the vehicle was buoyant and so made less weight on the tires. The tires kept spinning. I could feel them rolling cobbles as the truck groped for traction. Suddenly the grip improved as we came to a sandy part, and we were again making some progress. But the muddy water now swelled over the hood and up onto the windshield. Without warning, the front of the vehicle dropped as we drove over a small underwater ledge. We were in the middle of the river, and the hood was underwater.

"Completely waterproof," Leon said, smiling, glancing at me for only a second before returning his attention to the river. "But there is a big hole around here somewhere I want to avoid."

He avoided it, I guess, because we crossed and powered up a short, steep bank, stopping on the edge of a cornfield. A friend from childhood, Leon was eager to show me the powers of his truck. It had been a military truck. I don't know where he found it. There was no dashboard.

"Locking differentials, front and rear. I can disconnect the sway bars—just pull that cotter pin. And completely resistant to an EMP. No computer at all."

For Leon, an EMP, an electromagnetic pulse, was around every corner. An EMP might come from a nuclear explosion, or a solar flare. EMPs can knock out electronics, and popular thinking holds that were it to happen, it would render modern cars useless. To hear him talk about it, EMPs were one of the great threats to the American Way of Life. Some studies suggest that EMPs have very little effect on cars, and almost all the effects are minor and temporary. I didn't bring that up, though. Leon loved fantasizing about the end of the world as we know it. He was a prepper, I suppose, although most of his prepping involved toys like this truck that he probably wanted anyway.

"When shit hits the fan, you get up here with me. We'll be fine." He beamed as he offered this. In his thinking, the next apocalypse would be the best thing that could happen for him, and I understood. Leon had been an outstanding athlete in high school, smart but not all that interested in school. We rode dirt bikes in the mountains together. He graduated, somehow, and then worked for the gas company until a shoulder injury led to an opioid addiction. He had not really worked in years, and I think he still had substance-abuse problems. For him, any apocalyptic event would improve his lot. Suddenly, his skills would be valued, his strange obsessions converted to clever foresight. He would be an important person to know. I know the details of his apocalyptic fantasy because he talked about it all the time.

"I'm serious. You get up here as fast as you can. That's a two-hundred-foot well right there, and this whole river bottom will grow more than a hundred people could eat. And there are more deer than you can dodge with your car."

I had taken him to Honduras, to the rainforest, with me once. In the next apocalypse, he could finally pay me back. For Leon, the apocalypse would be a cure for poverty and addiction. It might even help with the self-loathing that can accompany that. In the next apocalypse, he would be a valuable member of the group. He did not feel that way now. Leon was waiting for everything to collapse, and his tough situation made that understandable. Even some of my college students, in far less dire situations, imagine the next apocalypse as a potentially positive redirection.

We have all thought about it. What would we do if everything came crashing down? We create varieties of narratives—some clearly fiction and some presented as our best guess about the future. The narratives we create become the reality we expect. These stories tell us a lot about ourselves, including what we want now, and what we hope for and fear in the future. Today, we seem to have reached new heights in the production of apocalyptic and dystopian narratives. Even a cursory examination of the apocalyptic media available to us reveals hundreds of films and thousands of books about dystopian futures. These are so popular that when I retitle my wilderness survival course as a "postapocalyptic survival" course, I get twice the interest. It's been termed "apoca-tainment" by Gwendolyn Foster.[1]

Media representations of the apocalypse certainly generate enthusiasm, but they can also limit the parameters of our thinking. Discourse matters, and everything from our vocabulary to the topics we choose to focus on can shape how we think about something, or even how we are capable of imagining it. The threats and fears presented in apocalyptic narratives are metaphorical representations of tensions that exist in the real world. From the critiques of racial justice to the xenophobia that underlies the narratives, nothing is merely about zombies, or a comet. The fear is not emanating from a virus, or a natural disaster, or at least not only from that. We see this play out in our recent experience with a pandemic. Our reaction to Covid-19 reflected ongoing political and cultural tensions, and the pandemic became a canvas painted by this struggle. As in the fictional apocalyptic narratives, the immediate threat became a cipher for an underlying concern.

There is a dark side to some of these fantasies. In some cases, the rhetoric that accompanies apocalyptic images promises a return to a traditional way of life, which sounds positive and conjures up wholesome images of satisfying, preindustrial, rural family life where hard work pays off. Of course, in the United States, that reality existed only for some groups. For most, the misogyny, racism, homophobia, and other "traditional" attitudes would make a return to the past overwhelmingly negative. The *status quo ante* of tradition is a more toxic version of the *status quo*, especially for those not protected by privilege. While broader contemporary society understands these ideas as backward and bigoted, a postapocalyptic world offers the opportunity to embrace them. These narratives inform how we think about the past, present, and future, and importantly, they influence how we act.

I am not conducting an exhaustive survey of apocalyptic literature here. The examples I discuss in the upcoming pages are ones that resonated with me as good examples of the kind of apocalyptic stories that I see as shaping our vision of the future. A few contemporary apocalyptic narratives stand out to me, either because of their place in the history of the genre (the book *Lucifer's Hammer*, or the film *Night of the Living Dead*) or because they embody certain approaches or points of view (the book *One Second After*). There are a few that stand out as masterfully artful examples of the genre, such as Cormac McCarthy's novel *The Road*, N. K. Jemisin's novel *The Fifth Season*, or the film *Mad Max: Fury Road*. There will be exceptions to any trend I identify, and I do not claim that the tropes I highlight occur in some particular percentage of narratives out there. In fact, that does not matter here. I am interested in the ones that make their way from narrative to real life, either in our actions or in our imaginations.

There are thousands of apocalyptic narratives. I am familiar with many of them, like most of us are, and I thought I had a sense of what was out there. I did not. I had barely scratched the surface. Some narratives paint a bleak and awful picture, like McCarthy's *The Road*, in which the protagonist fights an impossible battle to shield his young son from rampant cannibalism, cruelty, and despair amidst a dead world. Michael Haneke's *The Time of the Wolf* presents a similarly dark vision of the postapocalyptic world, in which a French family finds its potential safe-haven in their country home already claimed by hostile strangers, and after finding no help, and with nowhere to go, they wait on a train that might take them away from the chaos. Nobody would want those futures. They are bleak, hopeless, and lacking in compassion.

In many other cases, it is apparent that the thought of an apocalypse appeals to us on some level. Something about that imagined reality resonates with us, and we want some of what it offers. Perhaps this mirrors our experience with war movies, in which we present the hellish reality of war as an adventure story, a heroic epic. Perhaps we do the same to "the apocalypse," sanitizing and romanticizing something that is inherently awful. A radical change, though, may not be inherently awful. Some things need to change, certainly. Perhaps the apocalypse becomes shorthand for starting over and shedding the burdens we have accumulated.

One thing is clear: future apocalyptic scenarios are not presented in the same way as the disasters that we actually experience. There is little appeal to the aftermath of a tornado, or a house fire. Our apocalyptic fantasies, however, alternately horrify and attract us. I cannot explain away the appeal as mere schadenfreude, or as the kind of perverse pleasure we get from watching figurative train wrecks. Rather, our apocalyptic fantasies capture something we long for: the chance to do it all over, to simplify, or to get out from under something like debt or loneliness or dissatisfaction. It is decluttering on a grand scale. It allows the possibility of living life on our own terms. We can be heroic and put all of our skills to work. We can set our own agenda in ways that we currently cannot. We realize it would be tough, but we would be focused. Life would be hard but simple and satisfying. We tell ourselves that, at least. Many apocalyptic narratives reflect these fantasies, in which we can be the kind of hero we cannot be in our current lives.

In those instances where we seem truly terrified at the prospect of everything falling apart, we transform our real, visceral fears, via metaphor and proxy, into something else. That something else, that particular fictional apocalyptic future, manifests in a form where we can let go and fully express the terror that we might otherwise suppress and sublimate, in order to carry on in the real world. Those of whom we are fearful become zombies. The realities of a pandemic, with the boredom of isolating, the economic impacts and the constant, low-level fear is replaced by a rapid, deadly menace which requires heroic action to escape as protagonists drop everything and flee. In everyday life, we compartmentalize the fear, as we did with the fear of nuclear attack during the Cold War. In some of these narratives, we release the fear in a cathartic confrontation with the unthinkable.

These narratives are powerful. They inform how we act and react today as well as setting up a vision for a possible future. I see it when I teach my wilderness survival courses, as students' questions and comments about zombies or EMPs reflect the content of our fictional representations. I see it as I write this, during the Covid-19 pandemic. We struggled to behave in the correct ways during this crisis. People could not stay put or follow directions. We politicized medical advice and best practices. Part of this difficulty goes beyond emulating apocalyptic narratives and has to do with our celebration of individuality and self-sufficiency. These values form our origin stories, our mythical past, and shape how we respond in an emergency. Another part of our difficulty in managing the pandemic comes from the narratives we create of world-changing catastrophes. What we needed to do to survive did not resemble the behavior seen in apocalypse stories. In some ways, the pandemic was an anti-apocalypse, at least in terms of what we imagine one would be like. No one fantasizes about an apocalypse where you have to stay home and do nothing: your skills, your gear, and your preparations making little difference unless you had stocked up on toilet paper and hand sanitizer. But inaction, doing nothing, helped us survive. Staying put. Following the advice of experts. Compared to the typical apocalyptic narrative, the Covid-19 pandemic must have been the most unsatisfying apocalypse imaginable. Poet Bianca Spriggs called it a "domestic apocalypse," and I think that captures it in many ways.[2] We associate the word "domestic" with women, in keeping with our traditional divisions of labor. The domestic sphere has always been radically undervalued in our patriarchal systems when compared to public life. We labor under the illusion that certain differences exist between male and female brains and bodies that make one gender more suitable for certain roles. Public actions, individual heroics, in our patriarchal myopia, are the territory of men. But this time, heroics did not save the day: rather, most people operated as members of a community, came together to protect themselves, and that is what ushered us through. It hardly sounds like a bestselling apocalyptic novel. There are not a lot of novels about the 1918 influenza pandemic either.[3]

To understand the disjunction between our expectations and reality, I turn to the images of the apocalypse that shape our expectations and fuel our imagination. The recurring tropes provide insight into what we want and what we fear from the future, and there are several I see repeatedly in both popular media that we consume and in the manner in which we talk about the future while preparing for it.

The first and dominant trope of our apocalyptic narratives involves heroes and heroism. Our vision of meeting future challenges involves an individual hero, or heroic protagonist. In some ways this is understandable, given the way we construct narratives for books, TV, and movies. Even stories that focus on multiple groups at different places and times, such as *Lucifer's Hammer* (a novel in which a comet hits the Earth) tend to have individual heroic protagonists within each group. The fantasy of becoming a hero is not limited to apocalyptic narratives, of course, but it is strong and prevalent there. Apocalyptic fantasies allow a setting in which you can save your family or save the world, where you can use skills that are relatively useless (or unused) now, like some of the primitive wilderness skills that I teach, or skills against which we have societal prohibitions, such as the use of violence.

This trope has a couple of implications. First, the focus on overcoming the situation as a small group or individual means the larger community might constitute a problem. The second implication is that solutions come through heroic action, which takes great effort but is over quickly. Long-term solutions that do not show immediate results, or that are imperfect and always changing, are never part of the story. In reality, however, almost all significant problems require long-term solutions, and they are incomplete, perpetually adapting, and never finished. The immediate gratification of heroic action is missing.

I see evidence of the elevation of individual heroics over communal action when I look at the protagonists in popular apocalyptic narratives. Some are more overtly and classically heroic, like Brad Pitt's character Gerry Lane in the film adaption of *World War Z*, while others are less heroic in an "action hero" sense, such as the father in McCarthy's *The Road*. John Matherson, of

One Second After, best represents the heroic protagonist fantasy. Here Matherson parallels the author (ex-military turned historian at a small college in North Carolina), providing a remarkably transparent window into the author's fantasies. The author demonstrates the hero he imagines he himself could be in that situation. In that novel, the hero's principal task is to save his daughter—exactly the type of paternalistic masculine fantasy that permeates our apocalyptic imagination.

The heroic desire exists in the myths we create about our own history. In my case, myths featuring the essential Kentuckian or the pioneer (we do not say colonizer) in the preindustrial United States show up in popular apocalyptic narratives as the rugged, self-reliant individual who overcomes obstacles and creates a life for himself (or rarely, herself). In the United States, this type of rugged individualism forms the backbone of our collective aspirations and undergirds our national identity. The image of the "rugged individual" has deep roots in the self-sufficient frontiersman from the early decades of this nation, with that particular phrase itself dating to the 1920s. In fact, Herbert Hoover campaigned with that phrase in 1928 against the Democratic candidate, Al Smith, and promoted a type of laissez-faire capitalism, not unlike today's neoliberal capitalism.

A focus on the individual takes focus away from the community. When we are looking for solutions, using our narratives as a guide, we look for a hero to save us. This suggests that a heroic idea, a heroic personality, or heroic vision or courage will be required to succeed. Anybody who has worked in an organization or institution has seen a flashy personality, thirsty for glory, push to the front, with sometimes disastrous results. The quiet, thoughtful leader whose contributions become clearer in retrospect provides a better model for success.

In addition to celebrating self-serving or narcissistic personalities, the focus on the individual hero diverts our attention from the real problem. We fail to see the structural or systemic issues at play by making it an individual's problem to solve. Instead of fixing a broken system, we concentrate on finding the right hero. We think we can root out the bad cops, rather than considering that the system will continue to create them, ensuring the bad outcomes we wish to avoid. We put the onus of cleaning up our planet on the individual, with admonitions to recycle and not to litter, even though the much greater problem is industrial pollution and the use of resources in industry. As we will see going forward, many societal problems have no individual solution. You cannot provide food for yourself as part of a large group through a variety of individual efforts.

I grew up with heroes, including rebels and mavericks who embodied not only the self-reliant individual, but who broke the rules to circumvent an unjust or unwieldy system. In my family, we tell the story of Stiller Bill, an ancestor who made moonshine, and whose antiestablishment, outlaw image made me feel part of something conspiratorial and special. I had another ancestor, Miller Bill, who was much more successful as the owner of a grist mill. I cannot recall a single story about him.

My grandfather, Joe Begley, was a real-life hero to me. His environmental and social-justice activism from the 1960s until the day of his death in 2000 is well known in Appalachia, and he became something of a celebrity due to his actions, captivating personality, and Lincolnesque appearance. One often-recounted story recalls the bitterly cold day he stopped a coal train to keep people in his community from freezing. That cold, snowy winter, many vulnerable people were unable to replenish their coal stores, which was the heating fuel used in the pot-belly stoves that heated many Appalachian houses at the time. Folks were pulling up boards from their porches to try to heat their homes. People, especially old people, were in real danger. My grandfather called the railroad company and asked them to stop in town and dump some coal out onto a siding. He would deliver it in his truck to folks who needed it. The railroad refused, so he pulled his truck across the tracks, held up his deputy sheriff badge, and stopped the train, and then delivered the coal. To me, he was Robin Hood, Batman, and Abe Lincoln all in one. He was heroic.

I understand the attraction to heroes who courageously stand up to a corrupt system to do what's right, especially what is necessary for your community. Sometimes it works, as in that story of my grandfather, or in cases like Rosa Parks, Erin Brockovich, or Nelson Mandela. In all of those cases, however, the hero worked for the community. In contemporary apocalyptic narratives, by contrast, the community is reduced to the immediate family or small group, and the force against which the hero fights is not an unjust or inequitable system.

The kind of heroism seen in fictional narratives is reflected in how we prepare for the next apocalypse, and we can see that both the heroes in the stories and the ones we try to become have roots in a mythical past that we are attempting to re-create. Roxanne Dunbar-Ortiz effectively unpacks the mythical, heroic origin story of the United States in her book *Loaded: A Disarming History of the Second Amendment*. In particular, her take on the idea of an archetypal "hunter and his gun" image is useful to our discussion. The hunter is the outdoorsman, self-sufficient and at home in both "civilized" society and the wilderness, and has been accepted (even admired) by Native Americans, whose admiration apparently mattered even as they were dehumanized as either "noble savages" or uncivilized savages. She writes about James Fenimore Cooper's Leatherstocking Tales series from the 1820s through 1840s and the Daniel Boone legend created largely by 1780s publications by John Filson. She notes that "neither Filson nor Cooper created that reality. Rather, they created the mythological narratives that captured the experience and imagination of the Anglo-American settler, stories that were surely instrumental in sanitizing accountability for the atrocities related to genocide, and set the narrative pattern for future US writers, poets, and historians."[4]

The reality of the hunter, including the archetype of Daniel Boone, diverges from Filson's presentation. Rather than the almost Native American persona that was promoted, Boone was a commercial hunter, collecting hundreds of deerskins and other pelts to sell. Dunbar-Ortiz summarizes, "the legend and lore that mushroomed around Daniel Boone advanced notions of the hero explorer and adventurous hunter, and were written over the fact that he was a merchant, a trader, a land speculator, and a failed businessman."[5]

Where there is fear and trepidation, there is the chance to behave heroically. While many people dread the dramatic changes to come, others welcome it, itching for the chance to play the hero, and to use all the cool survival gear they have accumulated. The image of the heroic patriarch saving his family through his cleverness, training, and preparation fills the survivalist and prepper magazines on the newsstand. This type of heroic protagonist features heavily in our imagination of the future because it is so common in our narratives, and it's through stories that we understand and order the world. This narrative bias dictates the type of apocalyptic narratives we consume, but also those we create. In our prepping, we set our hopes on dramatic, individual action that will resolve a disaster situation. It will not.

Self-sufficiency is also an essential trope in our apocalypticism. Be prepared: the Boy Scouts' motto, and a creed for the imagined societal collapse that awaits us. Preparation takes on an almost moral tone in apocalyptic narratives, as well as in the prepper culture that I discuss more in the next chapter. Idiots and irresponsible people do not prepare, the reasoning goes. It follows, then, that the difficulty they experience is their own fault. This moral failing on their part absolves the prepper or survivalist of any responsibility or guilt for refusing to help them, because they committed the sin of not preparing. This, of course, mirrors the rhetoric used in politics when discussing poor people, or any marginalized group, really. Rather than recognizing that some people cannot prepare for a potential problem in the future because they are struggling with current problems, we label it an individual failing. As we often do, we can shift blame onto individuals for systemic problems. We erase historical inequities, and our complicity, when we blame the victim. We demonize those who do not achieve in a certain way, and we convince ourselves it is their fault. Since they brought it on themselves, we cannot be obligated to help them. Any assistance we do offer is above and beyond our duty, and makes us virtuous and praiseworthy. Our vision of the postapocalyptic world shares this with our contemporary world. In both cases, the narratives we create shape a myopic understanding of the world, present and future. They also shape our ideas about how to survive. We believe that through knowledge and preparation, we can "save" ourselves from a catastrophe. This may be true for a limited time. Before long, however, our task becomes larger, to re-create the structures and systems that are critical, such as agriculture. Our preparation needs to include the knowledge and training to re-create these systems, and that is a community-wide task. This is not to suggest that it is irrational or a waste of energy to train in emergency preparedness or other survival skills. But we need to understand these skills as short-term solutions that allow us to persevere long enough to do the real work as a community.

The idea that a life after a collapse could be simpler is among the more common tropes in our fantasies. It amounts to a global decluttering, removing many of the complications of modern life. We might envision this chance to simplify life as cleaning house. In many narratives, in films like *The Survivalist*, TV series like *The Rain*, and books like Cormac McCarthy's *The Road* or Emily

St. John Mandel's *Station Eleven*, apocalypse results in a life in which much of one's effort is dedicated to procuring basic necessities. Groups are very small, bureaucracy does not exist, and other people (outside of your group) are not part of your decision-making process. It is technologically simple and usually rural, or at least in an area with few other people. This simple life, or some version of it, resonates with many who are struggling with a complex social situation.

Sometimes, however, the complications that we seek to rid ourselves of may involve other people, including members of groups that we do not know, do not like, or do not want to accommodate. In some of these narratives, especially zombie narratives, this sort of cleaning house by the protagonists bears a resemblance to an ethnic or racial cleansing. In the narratives, the victims of the cleansing are fully dehumanized others (zombies) rather than the human victims in real life. If we look at common critiques of globalism, or if we look at the kinds of reactions to bilingual signs, such as the anger the "speak English in America" crowd feels at having to press "1" for English, we see a resistance to incorporating others into a group, for instance. We can see that this spring cleaning is a complicated process. The uglier side of the impulse for simplification and cleansing is not voiced but is evident nonetheless: eliminate the "other." We see this kind of sentiment in a variety of narratives, sometimes presented as the necessary destruction before rebuilding can begin. I saw it at the end of season 3 of *Westworld*, a popular TV series that debuted in 2016, when an apocalyptic event, planned and incited by the "hosts" (artificial humans) to destroy the humans who controlled them, was subsequently described as the necessary destruction of everything, in order to start anew and produce something better. The cleansing trope is appealing, but we immediately recognize the ways in which such a concept has an ugly side. Spring cleaning becomes spring cleansing.

Similar to this trope of cleaning house is the popular trope of choosing our own group after a collapse (all of these fantasies overlap). This variation on the simplification theme occurs in many stories, but you see it particularly strongly in narratives that feature people who have prepared for a disaster and are now dealing with some apocalyptic event alongside their chosen group. It is similar to how the prepper community envisions the apocalypse. *Lucifer's Hammer* engages this trope (when groups of survivors are selective in who they accept), as does *The Walking Dead* (with only certain people accepted within the in-group at the various survivor communities like Woodbury) and innumerable others. The construction of a compound into which we allow only selected people is perhaps the most obvious example of this type of impulse. We see it in many other ways though.

Sometimes we choose our in-group by allowing people in who possess some sort of particular skill. That is, a group bases acceptance upon one's usefulness or utility. This was an attitude that I saw with a speaker at a gun/survival show I attended in Louisville. One of the speakers mentioned that he and his group had a compound that was open to everybody in case of a disaster. He gave instructions on how to get there, which included meeting his security people down at the main road, and involved a sort of screening or vetting for utility. You had to have a skill deemed valuable in order to be admitted. This, of course, is part of the idea that everybody must pull their own weight in a disaster. The screening is not based on a set of unassailable criteria, of course. The skills that would be valued, or the attitude or demeanor that might be required, provide the opportunity to manipulate the makeup of the group. Certain groups of people are likely to have certain skills. If I made a rule that people of my group must be able to use a machete proficiently and know how to load a pack animal, I could guarantee that only people from certain parts of the developing world would qualify. Of course, that is not how it would work. Taken to its logical conclusion, this focus on utility reminds me of the "useless eater" concept from the dystopian alternative history series *The Man in the High Castle*, in which the Axis powers won World War II. The term designated people that did not contribute in acceptable ways to the Reich. That sentiment, or something close, exists just below the surface in some narratives.

Tests for usefulness form part of our vision of an apocalyptic future as part of the process of selecting your in-group, the people you will ride out the disaster with, and with whom you will rebuild. Sometimes, the question is posed very directly: *What do you bring to the table?* A real-life example from an unexpected source came during my discussion with film scholar Dr. Karen Ritzenhoff. She and I were talking about apocalyptic narratives when she told me about a recent blizzard during which her community in Connecticut lost power for several days. She went to her neighbors' house during the crisis and saw that they had a basement full of weapons and supplies. "I asked them jokingly if I would be in or out when the time came," she recalls. They looked at me, deadly serious, and said, 'What can you bring? Can you skin a deer? Can you shoot rabbits? Do you have any skills?'" The answer to Ritzenhoff's question, *Am I in or out?* became contingent on her utility. Not only utility, but utility defined in a certain way. Karen answered her neighbor, "'I can talk, I can teach, I can sing, I can paint. Would you need those folks too?' My lovely neighbors. They ruled me out." They had not read *Station Eleven*, apparently, where artistic endeavors are at the core of the story, which focuses on the survival of human culture as much as the survival of humans.

That test of utility, of course, is subjective. Immediately useful skills are valued, but the concept of usefulness can be manipulated in a way that ensures a certain kind of person will be selected. For instance, what if Ritzenhoff had answered that she could not hunt, but had experience scavenging? Would that have been seen as useful? Or what if she could pick locks and break into places? Both of those sets of skills could be immediately useful for survival but might be rejected because they are associated with certain groups, such as homeless people (in the case of scavenging) or thieves (breaking into places). Some people, like thieves, might be worth avoiding. In other cases, it becomes a convenient way to exercise our biases.

Sometimes, our vision of usefulness is limited. For instance, in Ritzenhoff's example, the teaching, singing, and painting was not valued. During the lockdowns related to the Covid-19 pandemic, we saw the importance of "nonessential" things like art and aesthetics taking a central place in our lives, with a renewed appreciation for movies, music, and conversation. Sometimes the usefulness of a skill is not as obvious as that of a hunter. What if a person had experience organizing people, or as a politician, or as a street-gang leader? When we interrogate the reasons behind accepting certain talents and rejecting others, we see that it is not an unbiased test of utility. It reflects whom they want, and what they want to become. Many of the skills are those associated with a rural life, and in the United States and other places, this carries implicit bias for the race, education level, and political orientation of individuals, as the rural/urban divide mirrors divisions in those other attributes.

When we choose our own group, we might simplify in a negative way. Fantasies often situate protagonists to avoid the messiness, discomfort, and difficulty of navigating a complex and fluid reality. Multiculturalism, or increased inclusivity and diversity, for example, can create fears of displacement, or aggravation at having to accommodate people not in your group. Such negative feelings are rarely manifested as openly racist or bigoted remarks. Rather, the celebration of certain backgrounds or attributes serve as proxy measures to weed others out. Celebrating a background that includes hunting, fishing, and farming, and creating a narrative where that background positions people to survive a catastrophe (the "country boy will survive" type of sentiment) serves to reinforce the whiteness of a group, as we associate this lifestyle with white rural America. There are many people of color in rural America, including many immigrants with rural, agricultural backgrounds, but our stereotypes do not include them, typically. This also encapsulates the urban/rural dichotomy that we see in many of these fantasies, in which urban areas are problematic and full of unprepared people, whereas rural areas produce people with "real" skills and innate survival advantages due to presumed experience navigating the natural world. The urban/rural distinction also carries racial undertones. When I was in college in Lexington, I heard a classmate's mother commenting that another student was "from Louisville," which is Kentucky's largest city. I understood, however, that she meant it as a euphemism for African American. These euphemisms, like "urban" or "inner city," allow people to talk about race without saying it, while maintaining plausible deniability. Likewise, certain descriptions of rural American upbringing are strongly associated with whiteness, much the way country music is imagined to be rural and white. "Urban" and "rural" have to be understood as ciphers for race. That connotation cannot be extracted from the next trope: that rural life is preferable to urban life.

The "cities are for suckers" mantra pops up over and over in prepper culture and in many fictional narratives. This urban-rural divide mirrors the liberal-conservative divide, and even has echoes of the type of Civil War discussions that I heard growing up—that Southerners were better shots, and that their rural hunting background made them better soldiers. This was repeated in WWI and WWII and continues to this day. The brother of one of my former students from a small Kentucky town was a soldier in Afghanistan during the wars after 9/11. He was interviewed for the TV news, and the segment concluded with the same tired trope of "country boy converts squirrel hunting acumen into great skill as soldier"—the same story we have been fed for 160 years, and integral to the rural-urban divide we see in contemporary prepper culture. A frequent comment on social media these days is that liberals do not have guns, and come the next civil war, conservatives will have the advantage. This echoes the decrying of the effete Easterner in the Old West that was a mainstay in cowboy movies of the mid-twentieth century. Liberals, at least white liberals, are imagined as prissy, city-dwelling elites, out of touch with the real Americans who live and work in rural areas. The "country boy will survive" trope might have arisen as a defensive response to marginalization by wealthier, more educated coastal dwellers, but it finds purchase in our narratives as a way to impart value to one segment of society at the expense of another. This plays into the red/blue political rift, of course, as that, too, is an urban/rural as well as racial divide.

In many apocalyptic narratives, survivors are family units or other small groups. The survivors are very limited in number, either because of global or local depopulation or because of some other kind of isolation from others. Individual efforts at agriculture (gardening, really) and some kind of neighborhood-watch type of security apparatus suffice for survival. One of the things these narrative choices accomplish is to further the illusion that small-scale individual efforts will somehow get us through the catastrophe. If I were the last person on Earth, food is not going to be nearly as big a problem as it would be if I were one of 100 million people whose industrial agriculture and food distribution systems have collapsed. If it is just me and my family surviving, I will not have to deal with disparate worldviews when designing solutions and I will not need to have the difficult and ongoing conversations and negotiations about how life will look when it is reorganized. My religious views, or even my political views, are unlikely to be a major source of conflict within a small family group that survives a disaster. My ideas about how people should behave, about gender roles, or sexuality, or which animals to eat, or how to raise children, are not going to be problems I face. English writer Brian Aldiss used the term the "cosy catastrophe" in a history of science fiction published in 1973.[6] He was referring to post-WWII British works in which an individual or small group of survivors of some catastrophe have an easy time of it in a landscape devoid of other people.

As the scale of these narratives is small, one of the things that is absent in the prepper fantasy is the resetting of large systems that could collapse, including the power grid, the food-distribution network, and large-scale farming. Rather, small-scale alternatives receive all the attention. On one hand, that is understandable, as it is the dependence on large-scale, complex technology that would have allowed the system to collapse in the first place. Therefore, resetting them as they were makes little sense. Failed power grids are one large system that might be successfully replaced with smaller-scale energy production or with the adoption of new or old technology that does not require the same type of energy input.

Part of the appeal of the smaller scale comes not from true utility but from the possibility that it would allow more individual and heroic efforts, unlike large-scale solutions. So, things like international supply chains are not the kind of systems that preppers talk about reestablishing. For one thing, it is beyond the scope of an individual or a family and so it might be left out for that reason alone. On the other hand, there is an element of isolationism at play in these communities, with a strong nationalist and xenophobic subtext. Just as Americans vilified China for the Covid-19 outbreak, or the way in which "Made in China" became shorthand for low quality, we tend to reject global connections in the small-scale future of our apocalyptic imagination.

The focus on family in these narratives typically involves a strong patriarchal bent. Apocalyptic narratives and prepper culture often promote a traditional masculinity, and many valued skills are presented as being part of this "lost" or "threatened" way of being. This struck me from my first foray into the world of survival instruction, and scholars have noted this as well. A recent study, looking at *Doomsday Preppers*, the popular TV program, concludes, "Male performances of disaster preparedness in reality television recuperate a preindustrial model of hegemonic masculinity by staging the plausible 'real world' conditions under which manly skills appear necessary for collective survival."[7] "Manly skills that appear necessary" is the key here. We see this in the attention paid to archaic, outdoor skills that require some degree of physical ability, like cutting firewood with an ax. People who depart from this traditional masculinity are portrayed as effete, weak, emotional (liberals) or incapable and fragile (city slickers), revealing the true nature of this masculine fantasy.

But these hypermasculine scenarios reflect no reality I have ever seen in all the time I have spent in areas where a kind of low-tech lifestyle exists. In my experience, traveling through the rainforest, our daily activities (gathering firewood, cooking, setting up and breaking down camp) were decidedly domestic—exactly what every woman in the village did every day. Back in the village, the men were usually walking to the fields, weeding and tending to plants, then walking home—hardly the kind of traditional masculine activity depicted in prepper narratives. Women gathered firewood, cut it, started fires, cooked, and performed general tasks around the village. I have gathered firewood for hundreds of fires. Never did I, nor anyone I observed, engage in the kind of ax wielding that Ronald Reagan famously displayed for photo-ops at his ranch in California (the Rancho del Cielo) as a shorthand for his masculinity, ruggedness, and vigor. Instead, most firewood was found as standing dead wood, or already fallen, and small enough to be easily broken by hand. I used a machete, the tropical tool that largely replaces an ax, nearly every day, but mostly for cutting weeds and almost never for firewood. Even if I had used it to cut wood, everybody uses machetes, and there is little association with gender. Axes are useful, of course, but depictions of their use reflect lumberjack fantasies more than typical activities performed in much of the world.

I notice this focus on traditional masculinity because the very same problematic attitudes permeate archaeology, my academic field. Archaeology struggles with a tendency to privilege scholars who do more fieldwork (as opposed to museum or laboratory work, for instance). We celebrate research in more rugged places. We continue the denigration and devaluing of the effete eastern seaboard elitist who does not spend their fair share of time in the field doing "manly" archaeological things. This dichotomy is captured by archaeologist Alfred Kidder in 1949, where he notes that "in popular belief, and unfortunately to some extent in fact, there are two sorts of archaeologists, the hairy-chested and the hairy-chinned."[8] This bespeaks a pervasive misogyny, in both archaeology and in our apocalyptic fantasies.

Many popular apocalyptic narratives reimagine the future as a mirror of an idealized mythical past, some sort of desirable way of life that is impossible now. It might feature a world without technology, a world with many fewer people, or a world where our concerns align with survival rather than the demands of everyday life in the complex, contemporary world. In the United States, this mythical past evokes the "Jeffersonian ideal" of an agrarian society. Of course, we can be nostalgic for an idyllic past that never existed. That is how nostalgia works.

Traditional ways of being, accompanied by traditional ways of thinking, can be racist, sexist, and rigid. Some of the sexism and racism in popular apocalyptic narratives is intentional and creates a horrific dystopia in works such as in Margaret Atwood's *The Handmaid's Tale*, where fertile women are enslaved and forced to bear children in the former United States, now a totalitarian theocracy called Gilead, following global drops in fertility rates and a civil war. Other times, as in the vision of the future in prepper culture, it reflects biases and desires that may not be voiced. One of the common ways in which sexist thought is perpetuated in these narratives is imagining that women need to be protected.[9] The protecting hero exhibits a kind of paternal masculinity, protecting the family. The idea that women are weaker and in need of protection relies on several assumptions about women's nature, tendencies, and capabilities that are problematic, if not simply false. Old-fashioned images of women as fragile, physically weak, emotional, nurturing, and nonconfrontational are demonstrably inaccurate. This should have always been obvious to anyone who watches a mother or sister take care of business, but it is the high-visibility, public demonstrations of female capability, from YPJ fighters in Syria to political leaders to women MMA fighters, that might ultimately undermine the sexist narrative.

Paternal masculinity is strong in apocalyptic stories. In her discussion of the *Walking Dead* television show, Katherine Sugg writes, "The apocalypse, especially as a television series, is a rather peculiar narrative vehicle for the articulation of a transformative future for—or a nostalgic return to—masculine agency and authority."[10] After reviewing so many apocalyptic stories, I would argue that the apocalypse is not a peculiar narrative vehicle. It is a predictable and even hackneyed vehicle for the display of a traditional form of masculinity. The "frontier myth" that permeates the series is common throughout many apocalyptic narratives because it embodies traditional masculinity, a landscape of heroic possibility, and a return to an uncluttered and technologically simpler time. This return to a real or mythical past sets the stage upon which traditional, often problematic, roles reemerge.

From a non-systematic glance at TV shows, magazines, and blogs by preppers, it seems to me that their target audience roughly parallels the kind of folks interested in Second Amendment rights—that is, white men. A closer look, however, suggests the culture is much more diverse than this. I found many examples of preppers of color, for example. They are more common than I imagined, but still relatively rare and probably not proportional to the population at large. There also appears to be a great range of socioeconomic status among preppers. For one thing, we know that the very rich, as they perhaps have always done, are preparing in case things fall apart. In my hometown of Lexington, a house in an affluent neighborhood came on the market complete with a survival bunker built into the basement. Many rural people who come from a range of socioeconomic statuses participate in prepper culture.

While there is more diversity than I imagined, the prevailing narrative seems directed at the majority. Some elements in the process of prepping would be interpreted differently, and encounter resistance, if performed by certain marginalized groups. Not only are the significant tasks involved in prepping easier with economic privilege, but some of these activities are easier for dominant groups to do than others. Certain groups of people are allowed to do certain things without raising any red flags, while others would be punished for the same activity. The ways in which white, rural preppers are able to arm themselves, construct defenses, and not engender some sort of legal, social, or other action directed against them would generate a negative reaction if another group behaved in similar ways. Compare the reception of the armed white militias by the police during the protests of 2020 compared to unarmed black protesters, or the official reactions during the Capitol riot of January 6, 2021. We are familiar with differential treatment for the same behavior, giving rise to concepts like "driving while black," or unequal punishment for similar offenses during the war on drugs. We have endless examples in which race, class, or nationality changes everything, and this extends to the kind of preparations that would be considered appropriate. For example, my cousins in rural Kentucky would not raise any eyebrows if any among them bought weapons, ammo, and military surplus trucks. An immigrant Muslim family might arouse a different reaction, however, and could run into trouble preparing in this way. This kind of disparity can even apply to things like buying property. If I bought some land in rural Kentucky, no one would think twice about it because my family comes from the general area and I fit right in with the demographics of most of the people who live around here. I could proceed to build a compound on that land with no real problem. At worst, I might be considered strange or eccentric. Imagine a situation where an immigrant family did the same thing. Actually, immigrants from Ireland would be fine. Immigrants from Iraq would not.

Related to issues of privilege is the question, "What constitutes an apocalypse, a disaster, or a catastrophe, and for whom is it a disaster?" The answer is different to different people. Sometimes, a loss may not be felt equally. The loss of the electric grid, for instance, was the sole catastrophe in the 2015 film *Into the Forest.* That same loss would not have constituted a catastrophe in rural Honduras. For some, the disruption of traffic constitutes a crisis worthy of violence, even if the impediment is part of a peaceful protest for social justice. The collapse of an economic system may not be a disaster at all for people who were not able to participate in that system. The Classic Maya collapse was not a catastrophe for neighboring groups. Some residents of former Soviet states may not consider the collapse of the Soviet Union to be catastrophic. Some collapses are liberating. Some are miracles.

Similarly, we can explore what constitutes an appropriate response to any given emergency or catastrophe. Is a dramatic, heroic, individual response effective and appropriate, or would a better response be a much slower, community-based response? One bizarre response to the Covid-19 pandemic has been the hoarding of guns and ammunition.[11] Some of the surge in gun sales occurred after the protests related to racial injustices and police brutality began in the summer of 2020, but some of it occurred beforehand, as soon as the scope of the pandemic became apparent. Perhaps this buying pattern was in response to expected shortages, and potential conflict caused by shortages. But the fantasy that people will quickly become unruly mobs, losing their humanity, is a common trope in apocalyptic narratives, and was likely one of the drivers in the uptick in sales.

What does it portend for the future that we are preparing for a sudden, dramatic apocalyptic event that demands heroic, individual action to save the day, and that validates the utility of traditional, paternalistic masculinity? What is lost in a future in which small groups rebuild at a small scale? Are we eschewing the possibility of cosmopolitanism in favor of clannishness through our elevation of a rural experience over an urban one? If this entire vision reflects privilege, and always already leaves somebody out, how do we adjust course? We inherited this vision of a future from our stories, and those narratives beget the next, and so forth. We conjured the apocalyptic narrative for which we prepare from the narratives we've already consumed.

As we saw in Part I, our historic and archaeological examples never played out in any way that resembles these apocalyptic fantasies. While remaining cognizant of the limitations of archaeological data, we see collapses happening slowly, over centuries, with a variety of causes and chain reactions. Cities fall, and empires disintegrate, but communities do not cease to exist. Agriculture continues, as does trade. The systems in place before continue in one form or another. Communities transform but persist. Heroic individual actions are not the sort of thing we can see from the archaeological data, but neither are they a component of the textual evidence for how events unfolded.

There are alternate visions of the apocalypse by authors who come from perspectives not typically represented in mainstream narratives. One interesting outlier in these apocalyptic narratives are those written by African American authors. These narratives often expose and sometimes invert the racism and discrimination that exists in the real world. N. K. Jemisin's magnificent Broken Earth trilogy, including *The Fifth Season*, provides a good example. While it is about a dystopian future where the Earth has been shattered, literally, by the ascendance of a group called orogenes, almost everything in the novel can be seen as a metaphor for the racial relations we see in the United States. Sometimes Jemisin inverts these relationships. The principal difference that separates humans is not race but rather the possession of the ability to control seismic properties of the planet. From the slang terms for those with these special abilities to the way they are disrespected in official encounters even if they are highly trained individuals, the parallels with the African American experience are clear, cleverly done and, of course, intentional. The text exposes these things metaphorically, allowing us to see the analogue with fresh eyes. While the protagonists here behave like those in other narratives, the nature of the fantasies and fears bear little resemblance to other apocalyptic literature.

CHAPTER 5

APOCALYPTIC FEARS

Susan Sontag, in her seminal essay "The Imagination of Disaster," masterfully interprets mid-twentieth-century science-fiction narratives. She sees us living under "continual threat of two equally fearful, but seemingly opposed, destinies: unremitting banality and inconceivable terror."[1] The narratives that we create allow us to cope with those threats. She outlines the role of science fiction, explaining that "one job that fantasy can do is to lift us out of the unbearably humdrum and to distract us from terrors, real or anticipated. . . . But another one of the things that fantasy can do is to normalize what is psychologically unbearable, thereby inuring us to it. In the one case, fantasy beautifies the world. In the other, it neutralizes it."[2]

Much of what she wrote applies today. We still discuss the Doomsday Clock, for instance, which was created in 1947 by the Bulletin of Atomic Scientists, and measures how close we are to self-inflicted destruction. We currently sit at 100 seconds until midnight (doomsday), downgraded from 3 minutes until midnight in 2015, and down from a high of 17 minutes until midnight in 1991, at the end of the Cold War. The Cold War, of course, shaped some of the most significant apocalyptic narratives, such as Stanley Kubrick's *Dr. Strangelove*. This significance derives from the artistry of the film, but also the timing and topic. In the Cold War, apocalyptic narratives for the first time referenced a real threat that could kill us all. World Wars I and II were horrific, and damaged our vision of humanity, technology, and the possibilities of the future. But narratives like *Dr. Strangelove* explored an existential threat we know to be real.

Legend has it that *The Godfather*, a fictional account of the mafia, inspired some people in the real mafia to reinvent themselves in Mario Puzo's image of them. We know that reporting on a particular phenomenon can create more instances of that thing happening. Apocalyptic narratives certainly reflect the fears and desires we have, but they can also create new fears and desires. One good example of this is the fear of an electromagnetic pulse (EMP), gleaned from novels like *One Second After*. Roscoe Bartlett, former congressional representative from Maryland, endorsed the book to the U.S. Congress, renewing interest in this phenomenon.[3] A TV movie from 1983, *The Day After* (a title that doubtlessly influenced the previous example), so impressed and disturbed President Ronald Reagan that it spurred him to seek ways to end nuclear proliferation.[4] The fear emerging from these fictional narratives can and has taken root in the real world.

Embedded in apocalyptic narratives are fears, or proxies of fears that we already possess. I sensed this from personal experience with my students, in an anecdotal way, but I spoke with researchers who study our fears of the future in a systematic manner. I spoke with writer Josephine Ferorelli and sociologist Meghan Kallman, founders of conceivablefuture.org, about the types of fears that people have, in the real world, that are related to the idea that a potentially catastrophic event looms in the near future. While the name of their project suggests a concern with procreation, Ferorelli suggests that once you begin to investigate the concerns, "you pull on the thread and it all comes unraveled." They soon realized that people's fears were complex, and they intersected with so many other concerns that it became impossible to untangle. They observed that 2018 saw the lowest U.S. birth rate in thirty-two years, and the rate had been declining four years in a row. Climate change topped the list of concerns, but others included debt, employment, housing, food security, and water security. These concerns are global.

The fear associated with apocalyptic fantasies reminds me of the themes that recur among preppers. There is a strong political component to the prepper world, and this fearfulness seems connected. Some of it is under the banner of advocacy for gun rights, or libertarian views, although prepping is strongly associated with right-wing political views, both overtly and more subtly. One of the best ways to see the connection between the political world and the prepper world is in some trends that occur alongside political changes. Prepper activity declined with the election of Donald Trump, probably because he calmed their fears.[5] Data describes the decline in the number of prepper trade shows, and the folding of those into gun shows. This suggests that the prepper activity was heavily partisan. Prepper-type activity increased among liberals since Trump's election, including a big increase in firearm purchases, mirroring what we saw with conservatives during the Obama administration. Maybe this type of partisanship is flexible and depends on external realities like which party is in power, and the perceived threats that come with not holding power.

Although examples exist all along the political spectrum, preppers exhibit a strong connection with conservative thinking, which embodies a "desire for order and stability, preference for gradual rather than revolutionary change, adherence to preexisting social norms, idealization of authority figures, punishment of deviants, and endorsement of social and economic inequality."[6] This was my impression, formed through casual exposure to prepper culture, but scholarly, systematic analyses reveal the same patterns.[7] There are exceptions to this, but the correlation is strong and unequivocal.[8] This matters because it relates to how people see threats and opportunities. It shapes our reaction to a crisis and can shape our efforts to rebuild after one. It matters because very few things in contemporary society demarcate and separate people as much as their position on the political spectrum, and political polarization is increasing globally.[9] The United States is now more polarized than the international average, and we feel this particularly acutely because we recently emerged from a time of lower-than-average polarization during the mid-twentieth century.

Worldwide, polarization is on the rise, but not everywhere. Polarization has increased in places like India, Brazil, and Turkey, but not in other places like Japan, Portugal, or Tunisia.[10] The press often describes this polarization as tribalism, and that language captures why it matters here. Political polarization can create a single point of "cleavage" that overrides all the other things that might unite people.[11] The result, when polarization is severe, is a schism that leads to each faction questioning the legitimacy of the other. In a crisis, I can imagine this tendency amplified to a point where it is one another's basic humanity we question. We see just such a thing historically in genocides and civil wars, from the Holocaust to the Balkan Wars of the 1990s. Polarization has clear and significant implications for how people prepare for the next apocalypse, and more important, for how they reconstitute themselves afterward. If preppers are disproportionately influential in creating our visions of readiness and post-disaster recovery, and if they cluster together at one of the political poles, the predilections and biases of that political outlook could color our view of how to prepare and rebuild.

Why does this association between preppers and conservative thought exist?* Preppers, by definition, are going to engage more closely with feared outcomes than people who are not readying themselves for them. Perhaps prepping attracts people who are more conservative because it fits into their worldview? Perhaps there are many ex-military and first responders among the prepping cohort, and these groups tends toward conservatism. Or maybe it's because of the conflation of survivalist/prepper culture with gun culture? In North America, at least, there is a high degree of overlap between prepping and self-defense, and that involves firearms. In the United States, few issues divide the left from the right like those surrounding guns.

A conservative mindset manifests particular psychological needs, including a need for order, a resistance to change, and a preference for traditional social hierarchies.[12] Recent research on the psychological and physiological differences between left and right suggests conservatives exhibit greater fear and disgust responses than liberals, and that conservatives see things as threatening and live in greater fear of those threats than more liberal people.[13] Liberals are less fearful, and perceive things as threats less often. Not fearing something shouldn't be confused with failing to identify a threat, however. Conservatives are not better at identifying threats than liberals are. They simply react with more fear to the same thing. Seeing threats where others do not might lead people to be part of the prepper world. Psychologists have suggested that while a conservative mindset might have been advantageous for humans when groups where small and interaction with other groups was limited, liberals possess a manner of thinking that works better for the contemporary, complex world.[14] This vision of the past presupposes a less complex world, where somehow we encountered fewer outsiders, which I find

* My use of the terms "liberal" and "conservative" to indicate position on the political spectrum might imply that we are fully one or the other, but we should understand them as a shorthand to indicate where the preponderance of our sentiments lie.

problematic. We find differences between factions, and we create outsiders, no matter how subtle or odd the defining attribute. We inhabit complex social systems, and we always have. Even small groups of people create enormously complex social worlds.* Perhaps a more fearful, conservative approach amounts to erring on the side of caution, useful when the costs of failing to identify a threat are greater than the costs of incorrectly perceiving something as threatening. Perhaps it serves to create a preference for people who are part of one's own group or clan, which would result in more effort on behalf of those sharing your genes.

Whatever the reason for the different reactions, fear is a stronger motivator among conservatives, as is maintaining order and continuing traditional ways. On some issues, however, the tables are turned. Conservatives demonstrate less fear of climate change than liberals do, for instance. The response to the Covid-19 pandemic suggests the way in which other factors complicate fear reactions, such as the cavalier reaction and mockery of safety precautions shown by the right wing. This *was* a case of failing or refusing to identify a threat. Conservatives do not simply react with less fear to climate change; they often do not believe it is real or, for political and other reasons, refuse to acknowledge it. There is no reason to think that a conservative viewpoint would be better at identifying potential threats in the complex, modern world, whether these are potential disasters stemming from climate change, unequal distribution of wealth, or racial inequality. Finally, the extra caution and fear exhibited by conservatives should not be confused with better situational awareness or envisioned as a mindset that increases chances of survival in a complex modern situation. In everyday discourse, conservative fear and trepidation can result in seeing threats that are not there. The next apocalypse will be complex and messy,

---

* Anybody from a small town will tell you that navigating the social world there, in a small group where everybody is connected, can be much more challenging than navigating the social world in a city.

and realizing our tendencies and biases will go a long way toward making us more adaptable, no matter where you fall on the political spectrum.

Looking at our preparations for the next apocalypse and at our narratives detailing our visions of it, several common fears feature prominently and parallel concerns in the present. In the past, this included concerns related to nuclear war, space, communism, and racial issues. Dahlia Schweitzer looks at outbreak narratives in her book *Going Viral: Zombies, Viruses, and the End of the World*.[15] I interviewed Schweitzer for a radio show that I host on our public radio station in central Kentucky called *Future Tense*, which focuses on our post-pandemic future.[16] We spoke about the pandemic and post-pandemic world. She sees the popularity of apocalypse narratives as being rooted in their ability to simplify moral ambiguities. We might be uncomfortable with other groups, for instance, but most of us realize we are not justified in judging or hating a group solely for our differences. If the "other" is infected with a dangerous virus, originating in another part of the world, however, we are justified in fearing and shunning them. During our conversation, Schweitzer noted that "viruses remain a powerful and infectious metaphor, a way to demarcate 'dangerous' people . . . a way to spread fear." She notes Donald Trump's 2015 declaration that "'tremendous infectious disease is pouring across the border' in the body of immigrants" as a cogent example of that kind of hateful othering. The reductionist simplification of a complex situation appeals to us in the way that simplistic us-versus-them rhetoric works in the political arena. People's tendency to simplify messy and complex realities shows up throughout my analysis of apocalyptic narratives.

For Schweitzer, outbreak narratives reveal far more about us than the storylines initially suggest. One of the things that interests her is the nature and origin of fear in these stories. "What is it that we fear, and what does that say about us?" Real-world worries make their way into these narratives, and in historical documents, the narratives reveal the prevailing fears at that time. These fears reflect tangible and dangerous things, like radiation, but they also indicate a generalized fear of the unknown. We see this during the advent of the space programs in the United States and Russia in the 1950s and 1960s. By the time we get to 1968's *Night of the Living Dead* and *The Andromeda Strain* a year later, fears change from the demonstrable threat of radiation to viruses from outer space. "The threat is up there," Schweitzer notes. The threat can be down here too. Placing the threat in space is a cipher for our fears of people we perceive to be fundamentally unlike us. The origin of the outbreaks in narratives reflects our fears of real diseases, but also our ignorance about unfamiliar people and places. The fear in the outbreak narratives in the 1980s and '90s reflected our fear of AIDS and revealed our colonialist image of "darkest Africa."

When you are a parent, one terrifying prospect is that you will be unable to protect your children when they need you, either in the present or the future. This fear permeates apocalyptic and prepper literature, as well as elsewhere in the real world. For a time, when I had been sick with malaria while we were expecting our third child, these fears became visceral. This manifested in many ways, but dreams were the most jarring. In one dream, which seemed real in the throes of it, I looked out the window of my daughter's room. A tornado was on the horizon, roaring and churning. Impossibly wide and violent, it was headed for our house. My vision narrowed, and my heart pounded. We would surely die if we did not leave immediately. My daughter was too young to walk far by herself, and my son in the next room was even younger. I had to carry them. But neither would wake up, and I could not get them both into my arms. I felt weak, clumsy, and impossibly slow. The roar of the tornado was deafening. Our time was running out, but I could do nothing. Suddenly, it was too late. I had failed, and there was no escape. The tornado screamed into the yard, and the terror escalated until the windows blew out in a maelstrom of glass and noise.

I jerked awake, scaring my wife. I breathed hard as my heart pounded, terror giving way to frustration that I could not get through one night without waking up with nightmares. I had been having them every night. It was not always the same dream, but it always meant the same thing: I could not protect my family. I had been having these nightmares for a couple of months as I recovered from malaria. I was skin and bones, still anemic and very weak. My dreams reflected that helplessness. After I awakened from the tornado dream, I lay there paralyzed by the depression that accompanied this debilitating disease, thinking how I could not carry even one of my kids to safety. At that moment, I was living part of my nightmare. Even before and after this, at some level, I had these simplistic and visceral thoughts about the minimal things I should be able to do to keep my family safe. One of those ideas was that I should be able to physically carry my kids, all of them at once, to safety. Protecting your family, in reality, rarely involves carrying them. Even if carrying were required, it seems insufficient to afford protection of any meaningful measure.

My dreams, expectations, and fears are all an example of the kind of paternalistic masculinity that permeates prepper culture. Among the prepper community, taking care of the family is paramount. That focus seems very obvious and natural (who doesn't want to take care of their family?), but the way in which it is presented demonstrates the underlying assumptions about who is responsible, capable, and in charge. Consider the dramatic photos that accompany some of the articles in prepper magazines. There we see this preoccupation with protection again and again. It can take many forms, but an image of a man standing in front of a woman (wife) and girl (daughter), placing himself between family and danger, is one common manifestation. The fear is for the family, but beyond this is also the fear of not being able to protect the family. Those seem similar, but the second includes not only the fear that something will happen to the family, but also the fear of being unable to fulfill the traditional masculine mandate of taking care of the family. This is a fear of failure, as much as a fear of the consequences of that failure. Popular apocalyptic narratives are full of this kind of paternal masculinity, which actually goes beyond the paternal urge to protect and includes an internalized desire to perform a certain style of masculinity. The small scale of this kind of situation—protagonist plus family—fits certain narratives well and might account for its popularity as a trope. That is, it's easier for movies or literature to have the number of protagonists you would find in most immediate families. Dozens, hundreds, or thousands of protagonists is hard to manage.

Fear of the "other" is one of the more common tropes found in popular apocalyptic narratives. I use the word "other" in the anthropological sense where we talk about "othering"—the process by which we imagine and amplify differences between our own group and some other, with the goal of creating a contrast in our favor. This necessarily reduces and dehumanizes the subject of the othering. In these apocalyptic genres, the dehumanization could not be more literal than what we see in zombie outbreak narratives, as Dahlia Schweitzer points out. Zombies are fully dehumanized others. In some other instances, the dehumanization takes the form of people who are behaving as animals, such as the cannibals in Cormac McCarthy's *The Road*. Sometimes the other is based on nationality, and nationalism as a potent form of xenophobia is evident in some narratives like *One Second After*, where the bad people who caused the apocalyptic event are North Koreans and Iranians, fulfilling a right-wing fantasy. In other cases, the othering is much more subtle. In *Lucifer's Hammer*, for instance, the "others" are those who did not prepare for the apocalypse and are thus deemed unworthy, treated as a problem and threat for not having done the proper things to protect themselves and their families.

Another common trope involves situations in which other people become the enemy. This overlaps with some of the aforementioned fears, of course, but here I mean specifically where the obstacle to overcome is people, rather than something like climate change or economic collapse. This trope involves situations in which, because of the realities of the apocalyptic setting, a group of people becomes a threat to your survival or stands between you and some solution. This often overlaps with the "preparation as virtue" trope I mentioned earlier, in which the enemy are the unprepared who are trying to steal your stuff. We see this scenario in so many narratives, from *Lucifer's Hammer* to the film *The Survivalist.* This might be a normal plot twist in an adventure narrative. Where you have a protagonist, you are almost always going to have antagonists. Remember back to high school where we learned the essential plot conflicts: we have person versus person, person versus nature, person versus self, and person versus society. In most of the likely catastrophic scenarios resulting in the next apocalypse, however, people are not the enemy; it is likely nature, albeit nature modified by us. Rather than person versus nature, it will be a community-versus-nature kind of struggle.

Related to the others-as-enemy trope are narratives in which people lose their humanity. In some cases, we dehumanize others, as in the above examples. In other instances, folks lose their humanity on their own and become savage. Even if we leave out fully dehumanized people like zombies, these narratives present situations in which you might need to protect yourself against rampaging gangs, cannibals, or criminals. In some, like *Lucifer's Hammer,* we see groups of people turning into homicidal looters within minutes of the comet striking the Earth. In *One Second After,* there are roving bands. In *The Road*, we see cannibalistic hordes. Even in zombie narratives like *The Walking Dead*, the danger from other people seems greater than the danger represented by the zombies. I keep coming back to the examples where preparation becomes moral and virtuous, much in the way Americans valorize hard work or cleanliness. This valorization of preparedness is seen in both fictional narratives and within the prepper community. Because there is a type of morality or righteousness to being prepared, the unprepared are deemed unworthy and deserve what they get. They become the enemy because of their desperation born of their own laziness and neglect.

This reminds me of the problematic and flawed vision of the world in which people are either wolves (predators), sheep (victims, unable to defend themselves), and sheepdogs ("hard men" who keep us safe, including soldiers and police). This vision of heroic, prepared sheepdog protectors defending the unsuspecting sheep from the wolves gained popularity as part of our valorization of the military and first responders after 9/11. The sheepdog mentality, voiced in such books as *Lone Survivor* and *American Sniper*, is a favorite in a world where the military, police, and first responders are seen as heroes, ex officio.

A loss of humanity for zombies or radicalized North Koreans allows violence to be acceptable, as these people have become an existential threat to the protagonists. We saw this play out in real life with the reactions to protests for social justice that were prevalent in the summer of 2020, where some of the rhetoric suggested that dangerous, rioting looters had become such a threat that armed vigilantes or gangs have a right, maybe even a duty, to protect themselves and their property from these kinds of dehumanized mobs. Framing others as having lost their humanity allows this kind of response.

One of the first apocalyptic narratives I read was *Lucifer's Hammer*, decades ago during my first archaeological field season in Honduras. One of the things I noticed in this book was how quickly people begin to engage in antisocial behavior once the comet strikes. Within minutes, one of the characters encounters a biker gang who have already murdered people, and vigilance becomes a critical skill. In the novel, I read the bikers' loss of humanity as partly stemming from character flaws, but also in relation to the righteousness of preparedness, since the lack of preparedness resulted in the antisocial behavior. This valorization of preparedness functions to create a "right" and "wrong" way of doing things and reminds me of the way that rigid parents might warn children against any deviation from a set of strict rules. The potential for people to lose their humanity, or even for them to become the enemy, means that you have to remain vigilant and recognize danger, which could come from anywhere. This focus on situational awareness is fundamental to the prepper community and makes its way into many fictional narratives as well.

The ideas we have about how people will behave in the face of a disaster are not supported by recent observations, however. Evidence suggests, as Rebecca Solnit documents in *A Paradise Born in Hell*, that people do not devolve into antisocial, problematic behavior in the aftermath of a disaster.[17] Rather, folks rise to the occasion.

Paralleling contemporary political rhetoric, one of our often-manifested fears is "losing our way of life." I am tempted to put a ™ after that phrase, given how much political rhetoric out there uses exactly that phrase. Preserving a "way of life" has become shorthand for rejecting the other, and reeks of racism and xenophobia these days. When we say "losing our way of life," we ostensibly mean our traditional ways of making a living, traditional geographical territory, language, or religion. When used in the context of a marginalized group facing a loss of language or land or traditional agricultural systems, it reads one way. When used as we typically see it in the political arena, and when we analyze its use as metaphor in apocalyptic narratives, it can instead be read as a fear of losing power and privilege, or a generalized dislike of the Other.

This sentiment is strong in some of our fictional narratives, either voiced by characters lamenting something lost, or searching for it. The loss can be quotidian things like electricity or a varied food supply, but sometimes the lament involves a higher order of loss, such as the disappearance of a society that supports specialized occupations such as artists, craftspeople, or teachers.

The fear of loss is used to another effect in the search for Twinkies by Woody Harrelson's character, Tallahassee, in the 2009 film *Zombieland*. A Twinkie, as a ludicrous object of desire, effectively conveys the degree to which loss extends far beyond the necessary, or even the good, to something largely rejected prior to the collapse and representative of the worst of industrial consumer culture. Even Twinkies, emblematic of the previous world order but with little inherent value, become a metaphor for a lost way of life. In other narratives, the loss is of something transcendent, like wisdom in the case of the religious texts in *The Book of Eli*, or knowledge in general, as with the forbidden books in the 2013 film *Oblivion*. We see the pain of the loss of a way of life in attempts to create a simulacrum of normality, from Hershel's farm in *The Walking Dead*, or even with all of the ersatz modern comforts on *Gilligan's Island*.

One attitude often reflected in these narratives is a discomfort with the idea of changing the roles we inhabit. We can see this in the fantasies that allow (or force?) us to return to traditional roles that are now outmoded, such as the paternal masculinity pervasive in the genre. We see people worrying over the loss of religion, or the shift in attitudes concerning what is decent or proper, or some sort of ethic that was valued previously but now holds no value.

In some cases, of course, the fear of losing a particular lifestyle or tradition is understandable and the pursuit of their preservation is laudable. Nobody faults an indigenous group for trying to preserve its language, for example, and nobody faults the historic preservation community for their efforts to preserve architecture or other materials from the past. On the other hand, like patriotism, the fight to preserve a "way of life" can serve to disguise racism, xenophobia, or sexism. Paternalistic behavior permeates popular apocalyptic narratives, but also most popular narratives in general, especially heroic narratives. Less directly, traditional ways of being are signaled in these narratives and in prepper/survivalist literature by a clear preference for skills (and even tools like axes) that have come to represent a vanishing traditional masculinity. Sometimes an ax is just an ax, I suppose, but not every time.

Violence is an assumed part of the reality of a postapocalyptic setting. The fact that the survival trade show I attended in Louisville was also a gun show is no accident. We see this when we look at the publications focused on prepping, such as the magazines that are now readily available in every supermarket. In addition to articles about wilderness skills or outdoor gear, you'll see articles about firearms that are focused on self-defense. Contrast this with other groups that are interested in some kind of off-the-grid existence that would be consistent with prepping, but focus on other aspects of it, like renewable energy, sustainable small farming, or community gardening. In those latter visions, there is no such focus on violence now or in the future.

Many visions of the future involve having to endure violence inflicted by others, but part of the fantasy for others seems to be the ability to use violence in ways that they cannot in polite society. In some of the prepper fantasies, the prospect of violence does not engender the reaction one would expect if it were merely a horrific part of a future reality. Instead, it seems to be presented differently. Sometimes, it strikes me more like an opportunity to use the equipment that they have amassed, to use violence with greater impunity. After all, if society has collapsed, you will not be indicted for manslaughter or murder, nor held accountable for a wrongful death. Prepper gun fantasies can be fully realized in this red-tape-free future. I see this as a subtext in articles discussing home protection, and it comes out in talking to folks about defending their food stores and equipment. Scenarios of using deadly force to protect your property, even if that property is your store of food, are complex, and might not be free of legal consequences.

Where real violence had affected lives, as I saw in Central America, fantasies are very different. They involve escaping violence altogether. Their fantasy is about not having to defend your home, rather than successfully defending it.

Stockpiling supplies and defending them are not part of my survival courses, which focus more on skills. Sometimes, though, my students ask me about guns, self-defense, and guarding your hoard of prepared materials. Personal self-defense has its place, but it's like the bushcraft skills I teach: they are sometimes necessary, for a limited time, but do not provide a viable long-term solution. They address a symptom of a problem, not the cause.

I have no problem with taking steps to defend yourself and your family, but I do fear that assumptions about who constitutes a threat will be directed by existing biases and prejudices. Sometimes, I sense that the focus on self-defense involves imaginary situations involving people who are already identified as a problem. A student in one of my courses asked about home defense, revealing his vision of the apocalypse. "There'll be hordes coming out of the city, into the suburbs [where he lives]. You have to protect your home." Yes, we should not be caught unprepared to resist an aggressor, I agreed, but I could not help but think that the characterization of the people streaming out of the city reveals a lot about the assumptions he was making, and who he saw as a likely threat. I also think we could frame this same situation completely differently. Rather than viewing people on the move as inexcusably unprepared (again, a moral failing in the eyes of some preppers) and potential looters, we could see them as people in need and as people with skills that might help us all adapt to the new reality. In the case of the more plausible catastrophes, there will be many people in need streaming out of affected areas. Rather than asking how to protect our stash of valuable goods from them, we could ask what these folks need, and how we can help. Incorporating new people into your group, rather than maintaining the lines of distinction, could be the safer and more plausible option. Of course, when that student described the situation he feared, I suspect he was envisioning "urban" folks streaming out of the city into the suburbs, with all the racial and ethnic implications that carries.

Guns are very much part of prepper culture. This plays into the Old West ethos and Second Amendment/right-wing gun rights advocacy. In some circles, the possession of firearms is as basic to preparedness as food or water. The arguments about guns and preparedness are difficult to untangle from the rest of the gun-culture debates, and this comes out in prepper culture. In prepper magazines, for instance, articles and ads that are "pro Second Amendment" or that extol the virtues of an armed populace are as prevalent as less political topics such as hunting rifles. For small scale, short-term protection, the arguments for arming yourself can make sense. However, an armed populace as a means to decrease violent crime on a societal level is not convincing. Most studies that have withstood scrutiny suggest that arming the citizenry does not decrease violent crime.[18]

Arguments focused on opposing government overstep, or tyrannical governments, are not telling the whole story, since what it purports to thwart is very specific. Invasive laws that impact women's choices about their own bodies are never targeted as "overstepping," for instance, nor are indefinite detentions of suspected (but not convicted) terrorists. In fact, some of the arguments seem to be after-the-fact rationalizations for unspoken desires. Maybe it is the desire to be armed against perceived threats beyond a tyrannical government. Maybe it is the visceral excitement of firearms, or maybe they become a tangible emblem of group identity: something to cling to and fight for in a system in which much of the right-wing base feels under attack. In some cases, they have been marginalized, dismissed as uneducated and uncouth, having little value even to the right-wing elite beyond the votes they provide. In other cases, the perceived marginalization is not real. As the popular saying goes, to those accustomed to privilege, equality feels like oppression.

Firearms may be for protection, for defense, but they are also for offense. Some prepper fantasies seem similar to the response of the right-wing "militia" members to recent protests. Both seem to reflect a desire to find a situation in which violence is an option. They go to protests, ostensibly to protect property, but few outside observers believe that their concern is really for that gas station or drugstore. They go to find purpose, perhaps, as protectors.[19] Others see the confrontation with protesters, the enemy, as the real goal.[20] They go with the hope of finding the opportunity to use the violence for which they've prepared and about which they've fantasized. In that same way, fantasies about violence that accompany visions of a postapocalyptic world are as much about the freedom and opportunity to use violence as it is defending yourself from it. The actions of the militias at the protests in 2020—traveling long distances to "protect property"—can be read as people traveling to confront the enemy. Liberals, Democrats, and people of color were vilified as un-American in the right-wing press. This is the type of delegitimization that accompanies political polarization and provides a glimpse of what we could see in an apocalyptic emergency.

Something often invoked in tandem with firearms is "situational awareness." This means a kind of hypervigilance where people train themselves to look for and identify threats. We all do that to some degree. We watch the weather, avoid a dangerous driver, or leave a bar when it gets too rowdy. We might heighten that awareness in certain circumstances—when traveling to unfamiliar places, for instance—looking out for people who might take advantage of an unsuspecting traveler, or for pickpockets in tourist areas. There are coherent, sophisticated, nuanced, and reasonable approaches to assessing one's surroundings. I teach a type of situational awareness in my wilderness skills course. One problem with all of this is that the identification of a threat will always be influenced by preconceived notions. We see threats in things that we have already decided are threatening.

I see a problem with a focus on situational awareness when practiced by untrained, bigoted, or poorly informed people, especially in places and situations where such threats almost never emerge, such as daily life in a small town, or at the grocery store. This harkens back to the now-familiar concept that people are either sheep, wolves, or sheepdogs.

Situational-awareness fantasies—looking for bad guys at the supermarket—are partly self-congratulatory. I have seen several nearly identical stories recounted on social media about seeing somebody whose posture and eye movements signaled that they were a trained sheepdog, not clueless and helpless like the rest of us. Of course, such mutual-admiration stories also serve to point out that the storyteller is just as vigilant, and thus can recognize the sheepdog quality in another superior individual.

That all sounds comical and childish, but this type of so-called hypervigilance results in people identifying threats that they have a priori defined as threatening. The shady characters who are identified as threats are those whose groups have already been identified as shady, and in the United States, this usually includes a strong racial component. Trayvon Martin comes to mind. The "Karen" phenomenon, of white women calling the police on Black people for innocent actions, such as jogging or entering their own house, is a manifestation of this misguided (and often racist) "situational awareness." Identifying threats that do not exist would be laughable if the consequences were not so significant. The shooting of unarmed Black people by police is a version of hypervigilance gone wrong, and of stereotypes and prejudice driving the identification of threats.

In both the "situational awareness" and the firearms discussions, I sense an attempt to create a geography of danger and of violence in a landscape where threats are frequently nonexistent. But heroic action demands a landscape with threats to identify and overcome. These fantasies seem similar to the war fantasies that many young people entertain prior to seeing war firsthand. With few exceptions, those who live through war do not glorify it, much the same way that folks who have lived through gun violence do not.

CHAPTER 6

READING AND WRITING THE APOCALYPSE

Popular apocalyptic narratives are everywhere, and we act them out in our preparations. As ubiquitous as they are, the kind of contemporary narratives that we see on the bookstore shelf or in the list of movie options are merely the latest iteration of a storytelling tradition that spans human history. Narratives about the end of the world are nothing new.

The oldest known apocalyptic narratives occur in religious contexts. In fact, the word "apocalypse" has come to mean the type of catastrophic event described in religious texts, despite an etymology that included a much broader meaning. Our ideas about the next apocalypse come from sources beyond contemporary popular narratives and prepper culture. Apocalyptic narratives are common in many religious traditions, and they can shape how we think about the past, the present, and the future in the same way as the popular narratives. Where I live, we associate apocalyptic ideas with evangelical Christians, a group that gained prominence as part of a backlash to societal changes in the 1960s. I first encountered evangelical Christianity and apocalyptic narratives in the late 1970s. I was visiting my grandparents in West Virginia, and the girl next door invited me to an evangelical Christian revival with her family. I had never heard of Armageddon, or the Book of Revelation. I knew little about the Bible outside of listening to sermons at our decidedly non-evangelical church a few times a year, and the occasional foray into Sunday school. I was much more interested in my new friend than the church, at least until they got to the part about Armageddon.

Suddenly, I was terrified. My eleven-year-old self thought, *How in the hell did no adult think to mention anything about this to me?* Concerned, I asked my parents about it when I got home. They both grew up in Appalachia, familiar with evangelical Christianity, but somehow they thought it a thing of the past. I still remember the words of comfort from my father: "If Jesus came back, he'd be a radical. We'll see him on the picket line, or riding around on a chopper with Hunter Thompson. You're in good shape." He pulled Thompson's *Hell's Angels* off the shelf and handed it to me. "There is more than one good book. Read this one." About five pages in, I had forgotten all about my impending doom.

Apocalyptic thinking was hardly just a thing of the past. It was also a thing of the future. This anecdote took place over forty years ago. Doomsday thinking was about to become much more mainstream. Today, it is still going strong, and shapes both our present and future. We can go back millennia and find stories that inform religious or cultural tradition that precede and predict the modern apocalyptic narrative. All across the Americas, for instance, indigenous cultures relate origin stories that mention the destruction of earlier worlds, previous iterations of the current one, representing failed or flawed attempts by deities to create the world in which we live. Maya origin myths that tell of various creations and destructions, for instance, include the Popol Vuh and Chilam Bilam, both ancient narratives, only written down after the arrival of Europeans in the fifteenth century but certainly reflective of earlier oral versions.[1] Twelve thousand kilometers to the east, in Mesopotamia, the Epic of Gilgamesh contains the story of an apocalyptic flood, and is at least four thousand years old, similar to the tale recounted in at least eight other flood stories from Mesopotamia, including the story of Noah that appears in religious texts of Judaism, Christianity, and Islam. Creation and destruction myths exist everywhere.

Religious apocalypses are not always eschatological (that is, concerned with the future end of humanity). Many incidents of destruction take place in the distant, primordial past, like the creation stories of many Native American groups that recount unsuccessful or imperfect worlds that were destroyed and replaced. Aztec creation narratives locate us in the Fifth Sun, or the fifth creation. The stories of destruction of the four previous worlds in cycles of creation and destruction is similar to what we see in many traditions.

Religious narratives differ in significant ways from the types of popular fictional narratives discussed earlier. First, many religious traditions include descriptions of catastrophes that occurred in a mythical, distant past as part of the cosmogony or origin story of a group. These events have already occurred, and thus have little impact on the present or future. They do not describe events that the listener or reader might have to live through. Nobody needs to prepare for Noah's flood.

A more significant difference between religious apocalyptic narratives and popular narratives is that religious narratives form part of a worldview and a belief system that is profoundly important to many people. Religious apocalyptic visions might be prophetic (for example, the Book of Revelation or the Apocalypse of John), or suggest fundamental differences between some primordial time and the present by recounting the eradication of one creation and the emergence of another. We see this kind of discourse, about the nature of creation, in the flood narratives that form part of indigenous traditions all across the Americas, from the Yanomamo of Brazil and Venezuela to the Maya of Central America to groups in the Pacific Northwest.[2] Secular apocalyptic narratives certainly have meaning, as evidenced by the effort we put forth to analyze how they reveal our fantasies and fears, but they do not explain the origins or fate of the cosmos.

The events in religious apocalyptic narratives have many and varied causes, and some of these are distinct from the popular narratives. Sometimes we see a catastrophic event resulting from unacceptable behavior on the part of humanity. The flood narratives from Mesopotamian traditions that continued into the Judeo-Christo-Islamic tradition recount punishment for bad behavior or disobedience. In the flood narratives of the Atra-Hasis (one of the early Mesopotamian epics), the king of the Gods, Enlil, becomes annoyed at the noise made by humans as their population becomes too great, and decides to destroy them with a flood.[3] A lesser god, Enki, warns the hero, Atrahasis, in a dream to construct a ship to escape. Of course, these are narratives that must be interpreted. Should we take the noise complaints literally? Is the noisiness a way to describe rebelliousness? Does the noise represent overpopulation, a theme in many Mesopotamian narratives of the era? In later versions of the Epic of Gilgamesh, Atrahasis is replaced with Utnapishtim and Enki with the god Ea, but otherwise recounts a similar plan to destroy humanity.[4] Humanity is similarly described as noisy, but the texts suggest the displeasure results from the people who had overstepped their place in the hierarchy by constructing impressive and loud cities, essentially behaving in a way not befitting their station. In the later biblical version, with Noah in the role of Atrahasis or Utnapishtim, the noise and self-satisfaction that angered the gods in the Mesopotamian narratives becomes wickedness, violence, and corruption. These religious narratives are cautionary tales, but do not prompt us to imagine surviving the event in the way that the contemporary, popular apocalyptic narratives do.

These religious narratives are important even though many are not prescriptions for how to survive post-apocalypse. They are instructive in other ways, though. Stories about a mythical past shape how we behave. Even secular stories of a mythical past can radically influence the present and future: we can see many ways in which the American "rugged individualism" narrative shapes our response to current events, and they establish our vision of the correct way to be, surely affecting how we will behave in the future.

Religious narratives that foretell future events, such as the Armageddon narrative from the biblical book of Revelation, also influence our behavior in ways that have tangible effects on the present and future. While there are examples from traditions around the globe, I am going to focus on my own tradition here: Christian narratives about the Apocalypse and "end times." These narratives describe a period of tribulations that precede the end times. During the time of tribulation, the duration of which varies among interpretations, people will suffer hardships and catastrophes, followed by the promised second coming of Jesus. We can imagine the similarities between surviving this time of tribulation and surviving the kinds of disasters imagined in secular contexts. Even in this case, however, the focus in not always on how to survive. A minority of Christians believe in a pre-tribulational rapture, in which believers rise bodily into the heavens before the trials and tribulations begin. In that case, then, the concern is how to be on the right side of what is happening and avoid the tribulations, not how to survive them.

Among certain groups of Christians, particularly evangelical Christians, belief in end times is strong, and many people see signs now of those times.[5] Matthew Sutton has written about this phenomenon in his book *American Apocalypse: A History of Modern Evangelicalism*. He documents the growth of these eschatological beliefs, from relatively marginalized to much more mainstream.[6] When eschatological thinking makes its way to policy-making circles in government, or into popular media in some other way, these narratives affect far more of us than just the adherents of a particular religious tradition.

These religious traditions impact people in a broader context. These apocalyptic beliefs have been important to the popularity of some evangelical groups, including their political success. A variety of terms describe the variations in Christian belief surrounding the return of Jesus. One term, premillennialism, refers to the belief that Jesus will return before a prophesied "golden age"—a thousand-year era of peace called the Millennium. Premillennial thinking also interprets literally the apocalyptic events described in the Book of Revelation. The rise of premillennialism dates to the early nineteenth century, and figures prominently in the origin of modern evangelicalism, which Sutton classifies as "dispensational premillennialism."* The "dispensational" part refers to the belief that history is divided into various ages, or dispensations, during which different schemes of stewardship and administration fall to humans. Sutton sees the belief in an imminent apocalypse to be key to understanding modern evangelicals, especially the political orientation and appeal of this group.

Some evangelicals form a large part of the Christian right wing, along with some Catholics and other groups. The Christian right emerged in the 1970s as a political force, in part as a backlash to civil-rights developments in the 1960s. This voting

* Dispensationalist Christianity can be contrasted to the more common federal or covenant theology. In dispensationalist thought, God charges humanity with being stewards of creation during different ages, or dispensations, using the appropriate guiding principles for this stewardship. We are in the Church or Grace Age currently, according to this schema. In the more common covenant theology, the covenant between God and people is the organizing principle.

bloc formed a powerful segment of the Republican base and helped elect Ronald Reagan in 1980. The prevalence of religious apocalyptic millenarian thinking has increased since the 1980s, paralleling an increase in evangelical Christianity and the Christian right since the Reagan presidency.[7] The impact of evangelicals on right-wing politics since the 1970s has received so much scholarly attention that the Organization of American Historians described it as a "cottage industry."[8] Newer research and journalistic work explores the role of such belief in more recent Republican administrations.[9]

In recent decades, we have seen the impact of the Christian right as it shaped environmental policy in the 1980s, and again currently. From Ronald Reagan's secretary of the interior, James Watt, arguing against environmentalism by citing biblical mandates to use the Earth's resources, to climate-change deniers citing the inevitability of the end of the world as a way to absolve us from any responsibility to take action, this particular vision of the future affects people far outside of evangelical or any other kind of Christianity. More recently Mike Pompeo, Secretary of State under Donald Trump, publicly invoked the Rapture (when Christian believers, dead and alive, supposedly will rise into the air to meet God) at political events, and both he and Vice President Mike Pence were part of a White House Bible study group led by Ralph Drollinger, known for his eschatological views.[10] This kind of thinking shapes policies and our path into the future.

I saw some of the political ramifications up close when I was a child. My father worked for the Department of the Interior, in the Office of Surface Mining. He and a team of lawyers enforced environmental regulations of surface mining, a method of extracting coal by removing the overlying mountainside or mountaintop, as opposed to more traditional mining via tunnels, as in underground or deep mining. Commonly called strip mining, surface mining became common after World War II. By the

1960s, it had generated a lot of criticism concerning its environmental and social impacts. Set up under President Jimmy Carter, the Office of Surface Mining (OSM) was formed as a regulatory agency. The creation of OSM was a meaningful victory for environmentalists, but the role of the office changed radically with the election of Ronald Reagan and his subsequent appointment of James Watt as secretary of the interior. Watt's policies were anti-environmentalist, and by his own admission, his religious vision shaped his views on the environment. There are many variations of him saying things such as, "My responsibility is to follow the Scriptures, which is to occupy the land until Jesus returns," or "After the last tree is felled, Christ will come back."[11] He believed that we needed only manage resources to last a few generations until Jesus returned, and he did not believe this would be very far in the future. His frequent use of the term "generations" reflects religious rhetoric surrounding this issue within his brand of dispensationalist Christianity, and is certainly biblical, but also suggests a relatively short time period. Just as you measure the distance between cities in kilometers not centimeters, using "generation" as the unit of measure rather than "centuries" or "millennia" suggests a short period of time.

For non-religious environmentalists, working to manage our resources in perpetuity, this was outrageous, not to mention a violation of the principles of a secular government in which one's religion should not dictate policy. Of course, Watt also said things like "I never use the words Democrats and Republicans. It's liberals and Americans," and also advocated using the "cartridge box" against environmentalists if the jury box and ballot box were ineffective.[12] This makes it difficult to know if his antienvironmental policies stemmed from his religious beliefs or simply his conservatism and desire to oppose anything associated with liberals, an association environmentalism continues to have. Even as a teenager, I remember great consternation among my father and his colleagues, and a fundamental change in the atmosphere at his office. While legislation takes time to effect change, the attitude of Watt and the new administration had immediate effects. Watt and his cronies withheld support for the office through many means, including not filling vacant positions, radically decreasing the number of mine inspections, and countless other small actions and inactions. This undermining of the mission of the Office of Surface Mining appears to have been driven in part by Watt's religious worldview. This in turn affected people in the contemporary world with no connection to his brand of Christianity. Water quality, biodiversity, and even the security of living downslope from surface mines all changed when Watt's religious worldview manifested itself in governmental responses. The environmental consequences of his policies and precedent continue to affect us.

The dispensationalist worldview has not disappeared. I investigated an allegation of a mining official citing religious doctrine as a rationale for policy while working as a newspaper reporter for the *Mountain Eagle* in Whitesburg, Kentucky, in the 1990s. A similar argument was recently expressed in an op-ed by a mining engineer in a Lexington newspaper.[13] And more broadly, outside of the concerns of environmental policy, anti-abortion sentiments and anti-LGBTQ initiative typically intertwine with the religious beliefs of conservative politicians.

During the coronavirus pandemic, Governor Andy Beshear of my home state of Kentucky won praise for his handling of the emergency, including his unvarnished reports on the impact of the virus in the commonwealth. He met some resistance from religious conservatives, for whom disobedience to the public health mandates had become a political statement. On Easter Sunday, 2020, predictably, some churches defied the orders against public gatherings and held in-person services. Governor Beshear sent the state police to take down license-plate numbers and mandate quarantine for the participants. These actions were decried as violations of religious freedom by religious groups, followed by claims of martyrdom and a public presentation of victimhood that has become enormously attractive to many conservative Christians. In news interviews of parishioners emerging from these in-person church services, churchgoers expressed the idea that they were not vulnerable to the virus because of their faith.[14] This was repeated time and again. Like with James Watt, we see the commingling of religious and political views. During the same set of interviews with people leaving the church service, the pastor warned the correspondent not to report "fake news" about him, a conservative political catchphrase that reveals the subtext of the interaction. This statement reminds us that none of this occurs in isolation. Predictably, cases of Covid-19 spiked after these gatherings and similar ones around the country, resulting in the death of several parishioners and pastors. A type of fatalism accompanies some of the millenarian embrace of apocalyptic "end times" narratives. There are certainly "avertive millenarians" who believe that the end can be avoided through proper action, but theirs is a minority position.[15] There is evidence that religious people, including those who believe in religious apocalyptic narratives, more readily believe stories about the end of the world.[16]

I see a few principal issues that connect religious narratives to behavior in the present and future. First, the posited short time span for the remainder of humanity has implications in how we use and manage resources and how we respond to climate change. This affects environmental policies, as we have seen, and these policies have real-world implications that shape the future of humanity.

Second, there is a behavioral or obedience-related element to some of these religious narratives. In this kind of worldview, what happens to us, both our earthly and otherworldly fate, depends in part on how we act. That has obvious implications for issues like diversity, religious freedom, and ideas of sexuality and gender. When you envision one particular way of being as being the "correct" way, mandated by tradition and religious doctrine, and enforced by the threat of punishment in the form of some apocalyptic event or eternal damnation, you have powerful forces combining to control behavior. In the face of a real catastrophe, this kind of prescriptive model has implications for laying blame and for limiting the parameters of an acceptable response. For instance, certain ideas about gender roles could limit the potential organization and response of a small group, or ideas about sexual behavior could affect housing arrangements or the ways in which people work together. I am reminded of Mike Pence's admission that he would not be alone with a woman other than his wife, a rule apparently common in evangelical circles.[17] Widely decried as sexist and discriminatory (as relationships with women colleagues, for example, under these circumstances could not be equal to those of male colleagues), this attitude is problematic in its implicit assumption that every interaction with a woman is fraught with the potential for improper sexual behavior. Again, the potential limitations on the ways in which a group might conduct itself are obvious.

We understand the world through narratives, and they shape our interpretations of the past and our images of the future. To a degree, our stories and our imagination determine what happen to us. They shape our reactions. They do not control everything, obviously. Changing the narrative will not stop the climate from changing or deflect an asteroid. From here, we move into the future, considering what might happen, what it would look like, and how we might make it through the next apocalypse.

PART III

The Future

The Next Apocalypse

Diving at 120 feet deep, you have less than fifteen minutes of bottom time before you have to surface. Through the clear Mediterranean waters, I saw the telltale curve of a ceramic amphora on the seafloor and went down to check it out. As I descended, I saw the shape repeated, over and over, in an area extending more than seventy feet in one direction, but only twenty in the other—the oval shape of an ancient cargo ship. The amphorae were covered in marine growth: the yellow sponges and red and gray coral camouflaged them. You find them by looking for the shape.

I had to work fast, taking photos and writing notes in a waterproof book, all the time checking my dive watch to make sure I did not pass the allowed time at this depth. I looked at the handles, noting from the shape that the amphorae were from North Africa, and nearly two thousand years old. They had made it this far, to the island of Fourni in Greece, but not to their original destination. I wondered about the crew. We were near enough to land that some of them could have survived, but if the ship sank in rough seas, I suspect none did.

Our team of archaeologists found forty-five shipwrecks around that small archipelago in Greece. As I was out on our zodiacs, diving on ships that had sunk over a thousand years ago, I was aware of the maritime disasters happening only miles away in the Mediterranean. Instead of professional merchant sailors, which I imagine crewed the wrecks I found, the current disasters involved refugees from the Middle East and Africa, fleeing conflicts and unlivable conditions. It was the same story that played out a thousand years ago in Guatemala, where warfare, climate change, and environmental degradation forced people to move. Migrating across the Mediterranean, an estimated thirty-four thousand people have died in the process at the time of writing. The migrant crisis in Europe reflects the past and anticipates the future.

As an archaeologist, I understand the past by looking at the traces of behavior encoded in the things we leave behind. As an anthropologist, when I look at the present, I can find clues to things left unsaid in the things we create. When I look toward the future, however, it's a little like trying to walk backward while looking in a mirror to see behind me. I am looking in the wrong direction; in this case, I am looking to the past, while in the present, to see into the future. It seems nonsensical. However, the past is infinitely larger than the present, and we have a much greater variety of experiences and situations from which to learn. This conjuring of the future from the past merits careful scrutiny. There are no crystal balls. That said, predictions are educated guesses, not made on a whim and not without evidence.

Grounded in where we are now and what has happened historically, I examine how the future might play out. I look at whether the next apocalypse has already started, and what the immediate effects might be. I discuss the likely long-term concerns, and what skills and knowledge we should acquire to make it through.

While an "apocalypse" often presumes a negative outcome, change, even radical change, can be positive. Societies collapse, throwing people into chaos. It is entirely plausible, however, that the changes could be liberating. As we saw when looking at the archaeological examples, we can shape our expectations through vocabulary, and change might be good for most people. During the Covid-19 pandemic, I've heard many people talk about a return to normal that would be a new, better normal, and not the same old, problematic reality that had been exposed through the pandemic. A whole series of anthropologists imagined post-pandemic realities that were positive.[1] I contributed a less positive essay to the conversation, but the hope for a new normal was everywhere.[2]

In order to know what to expect and how to react, I keep in mind the differences between the past and our fantasies of the future. I look at if, when, and how the next apocalypse might occur, what we can do to survive, and how the past might provide a map to navigate the future.

CHAPTER 7

LIKELY SCENARIOS

Pancho and I walked in silence along a muddy road, through a ghost town, in the Honduran rainforest.

"I built that school," he said as he gestured to an abandoned wooden building. The wooden shingles and hand-hewn plank walls looked much older than the thirty years or so since he built it. He also showed me the house he constructed for his young family.

We camped on the edge of the ghost town. I was building a fire to cook dinner when I saw him standing at the far end of the clearing, his back to me, looking at something. He just stood there, and he was still standing there once I had the fire going and the water on to boil.

I walked over to see what was so interesting. His head moved slightly as he heard me approach. He wiped his face with a handkerchief, turned away from me, and walked off without a word.

He had been looking at an enormous rosebush, and I knew what it was and why he walked away.

Pancho and his family had left this little village thirty years earlier, when violence arrived in the form of a powerful family set on claiming the whole valley. Rather than fight a war of attrition, Pancho and the others left. They left behind the school and the houses. Pancho also left behind a son. The rosebush he planted as a grave marker towered above me. I looked at the roses, then down at where the tiny grave must have been. I turned and saw Pancho standing over by the fire, this time looking off at nothing.

The personal impact of dramatic societal change will show up everywhere, in a thousand little ways, every day, as we remember life before, and the price we have paid. We cannot predict the particulars of these impacts, but we can see a preview of what they might look like. I see it in the faces of people at the grocery store, the stress and worry. Some of this visible tension is economic and political stress, certainly, but likely also a concern for the future. I see it with my college students who are not interested in committing to a long stretch in graduate school, or more dramatically in some who simply quit trying in their courses.

Some threatening scenarios are already happening, like climate change and pandemics, and others could still happen in the future. For years, many of us have imagined a climate-change-related apocalypse. In 2020, our focus shifted to the pandemic. The virus created immediate, tangible problems, understandable to most everybody (illness, death), whereas the effects of climate change, although becoming more and more obvious, are more subtle. To truly understand climate change, you have to analyze years' worth of data. Since few are able to do that, we have to listen to and trust experts. Varied responses around the globe to the pandemic reveals the degree to which the opinion of experts may be rejected. In some places, ways of knowing the world may not align with what the experts do, as in the case of certain religious groups. Sometimes, particular histories or ulterior motives undermine the trust in experts. In the United States, for instance, an underlying anti-intellectualism has led to a politically charged distrust of experts. Of course, denying climate change has been a convenient political position for decades.

As an archaeologist, I see evidence that a profound, perhaps apocalyptic, change is bound to happen sooner or later. There is a nascent field of study dealing with the likelihood and nature of societal disintegration, known as collapsology.[1] In a recent book, two French collapsologists, Pablo Servigne, an agronomist and biologist, and Raphaël Stevens, an "eco-adviser" who focuses on the resilience of socio-ecological systems, conclude that collapse is likely, but they argue that simplistic statements like this do not capture the complex future. They stress the need for change, suggesting that a belief that we can continue as before is akin to a utopian viewpoint, while realists recognize that a transition is coming. Looking at the collapsologists who assess our current condition and estimate the potential for widespread, profound change, there seems to be widespread agreement that cataclysmic change is a real possibility, maybe an inevitability, and climate change ranks up there as a likely cause.[2]

One of the larger-scale studies on ancient collapses comes from Luke Kemp, a researcher at the Centre for the Study of Existential Risk at the University of Cambridge. He writes about the chances and causes of collapses, past and future, defining collapse as "a rapid and enduring loss of population, identity and socio-economic complexity. Public services crumble and disorder ensues as government loses control of its monopoly on violence."[3] Kemp surveys several ancient civilizations and concludes that the average life span of a civilization is 336 years. He also comes up with several metrics that he suggests can help us understand the chances of a collapse occurring. Many of Kemp's conclusions are derived from the work of archaeologists, emphasizing not only proximate causes but the complex systems in which we live and work, and the ways in which those systems can fall apart.

Despite the difficulties inherent in prediction, we do have clear indicators of the way things are going, and we can estimate the ways in which certain things are going to change. Climate change, for instance, is well researched, and while there are many unknowns and new variables that could come into play, there are parameters, or levels of confidence, that we can use to talk about events we are confident will happen. We are not merely guessing about the future.

We have some idea of certain things that are going happen, but we might not know whether they are going to constitute an apocalypse or a catastrophe. Everything suggests that it is not a question of *if* but *when* these events will happen. As an archaeologist, I would say yes, they are going to happen. It always happens. In some ways, though, that is not the interesting question. I want to know when, how, what it will look like, and how it would affect different people. I also want to know if there is any way to prevent or to mitigate the things that are going to happen.

Civilizations do not last forever. What I am concerned about is how fast change will come, and how dramatic it will be. Judging from historic examples of apocalyptic events, the whole process could take a very long time, be multicausal, and be unevenly felt among the population and across space and time. I do not know exactly what the end of our civilization will look like and I wonder about whether it will be understood as a catastrophe. From our data about the past, I imagine the process of collapse has already started, with environmental problems as one cause and political and social issues—particularly inequities in wealth and power exacerbated by neoliberal policies over the last half century—as the other. How quickly the unraveling will proceed, and how long until we realize that the process is going on, are harder to divine. Some processes, like climate change, are understood sufficiently well that, while unknowns exist, evidence suggests it will create profound and negative changes on a global scale.

My next concern is how soon something is likely to happen. For the moment, to answer that question, let me set aside some of the unlikely scenarios that could presumably happen at any time, like a meteor strike, and talk about the likely scenarios, starting with climate change. In terms of how quickly climate change could bring about a catastrophe, most climate scientists suggest it is already happening. Many of us nonspecialists get the same impression. The hurricanes, the fires, the changes in rainfall patterns and other weather conditions that have affected agriculture might already constitute real catastrophes. How quickly could climate change become so drastic that it would constitute an apocalyptic event? Quicker than we think. There is some disparity in thought on this, but scientists routinely talk about truly significant changes in climate within the next fifty years, and many talk about catastrophic change in twenty or thirty years. In that case, it appears that we are talking about something within the lifetime of today's young people. Overall, I think we radically underestimate the speed and seriousness of the change we are beginning to witness.

Other causes for an apocalyptic event, such as a pandemic, can emerge quickly. In order to estimate how long we have before we can expect an apocalyptic pandemic, I can look at how often they have occurred in the past, what causes their occurrence, and whether I can expect such occurrences to accelerate in the future. In terms of pandemics, two things suggest that they will be happening faster. One is the loss of habitat for the animals that might be carrying the contagion for the next pandemic. Humans positioning themselves to be in closer contact with animals that were formerly more remote will certainly speed up the process and the opportunity for currently unknown diseases to pass from animal hosts to human hosts. The other factor would be increased population density. With more and more people on the Earth, the ability to contain outbreaks to a small number of people becomes more and more difficult. Our increased global connectivity will also increase the risk that an outbreak could become widespread.

Looking at the first two decades of the twenty-first century, during which we saw SARS, MERS, and Covid-19 emerge, it seems reasonable to expect something widespread every few years. However, the 1918 influenza and the 2020 Covid-19 outbreaks are the only two that became global threats. One thing I notice about pandemics is that they play out in ways that are both unpredictable and unnecessary. During the 2020 pandemic, slow and poor response greatly exacerbated the situation. Premature reopening of public spaces changed the course of both the 1918 and 2020 pandemics. The process of outbreak prediction is complex, and unanticipated elements, unrelated to the pathogen itself, can completely change the trajectory of a pandemic.

Previous collapses were processes that took a long time, as they did in the cases of both the Classic Maya and Western Roman Empire. If a collapse were to occur in the future, it seems likely that we would not recognize the beginning of the process until we look back, after the fact. In retrospect, we will see that the next apocalypse started long before we started considering the possibility of it, and certainly long before we realized it was happening. Something sudden, like a volcanic eruption or a meteor strike, would not follow this pattern, clearly, but what we see in the past and probably in the future are small changes that beget other changes. These begin to add up exponentially, until they create a significant shift. Belatedly, we will wrap our minds around the change that is already happening around us. The next apocalypse will sneak up on us. It might *seem* sudden, as there could be an epiphany when we realize what is happening. When we analyze it, we will find the process started long ago and progressed undetected or ignored for decades.

POTENTIAL CAUSES OF THE NEXT APOCALYPSE

I divide potential causes into ones that are likely and those that are less likely or unlikely. I am not sure "cause" is even the correct word here. As we have seen, none of the dramatic declines or transformations that have occurred historically had a single cause, even if a single thing set off the process. Rather, a web of connected and interdependent phenomena occur in a historically specific setting, yielding results particular to that reality. In this way, a drought in one context might result in significant change of one type, while a very similar drought might result in something very different elsewhere. This is what we see in the case of the Maya lowlands of Central America, for instance, where droughts at the beginning of the Classic period yielded very different results from a similar drought at the end of that period. When we talk about causes for the next apocalypse, we might need to think in terms of the proximate cause, or the thing that set off the whole process, rather than some sort of phenomenon that, alone, determined the course of history.

In the same way that we retrofit our current concerns to past events, we will be limited in our ability to imagine the future by parameters we may not even realize we have established.[4] We cannot imagine every eventuality. We are likely to think about the causes of the next apocalypse based on our exposure to narratives about apocalypses and media depictions of apocalypses. The reality might be significantly different, in terms of what is likely to create great change or even apocalyptic change in the future. The tendency to prepare for the last war, not for the next, could result in us failing to understand how things have changed and how those changes could influence what happens in a catastrophe.

Climate change is the disaster I see on the horizon. It is not a question of if it will constitute a disaster but when it will happen and how bad it will be. Climate change will affect different places in different ways, and some areas might be spared a dramatic change for longer than other places. The American tropics, for instance, are likely to suffer both stronger storms, including hurricanes, and increased drought. Where I live in Kentucky, in the eastern United States, ranks very high in terms of negative climate-change effects in this country, including extreme heat, drought, wildfires, and flooding.[5] Various points of no return are commonly cited as indicators of impending disaster, including the tipping point of an increase of 1.5 to 2 degrees Celsius, which we are rapidly approaching with little sign of the types of mitigating efforts required to avoid those milestones.[6] Many changes are already evident, from extreme heat to powerful storms to wandering polar vortices. The questions that remain are how soon the effects will be felt and how extreme they will be. For many of us, changes in climate and subsequent changes in agriculture and fishing comprise the clear and present danger. As demonstrated in the past, the decline in one system (agriculture, for instance) triggers a crisis in another system (economic, for instance) and ultimately more (political, for instance). This pattern of an initial disruption, followed by economic problems that beget political instability repeats everywhere, from our previously explored examples of the Classic Maya and the Western Roman Empire to modern examples such as Syria. In my view, this pattern will precipitate the next widespread, profound change.

We are constantly evaluating where we are with respect to having crossed some tipping point that will make it impossible for us to limit global warming from climate change to something that we can live with. Periodically we discover that it is much, much worse than we thought, and that something hitherto unforeseen, like melting permafrost, exacerbates the situation in ways we did not account for. In other cases, there is good news, and we see results that suggest we had overestimated or not factored in something that would be a positive force in limiting this change.

I see no evidence that the political will exists to do anything until it is too late. I suspect the next apocalypse will be set into motion by issues related to climate change. We already see climate refugees fleeing areas affected by hurricanes, and we are going to see it in areas where we have drought or crop failure. This movement of populations leads to all sorts of other problems, as we are seeing in the political realities in Europe, where they are dealing with one set of migrating people, and in North America, where we are dealing with another set. I expect political discord and conflict to arise from the types of disruptions created by a change in climate.

As I write, in proximity to the 2020 pandemic, the likelihood of a global outbreak creating significant disruption also seems high. In our interconnected world, a change in one place results in changes in others. This has always been true and is not a new phenomenon, although globalization over the last century has exacerbated the impact of direct interaction on the ways in which pathogens spread. We live in complex systems, and the different pieces of these systems interact with one another and feed off one another. It comes as no surprise to me that problems very quickly escalate and eclipse whatever it was that triggered the event in the first place. In other words, a drought or a hurricane might be the proximate cause of a catastrophe, but very quickly all elements of the system bear the stress of that drought or hurricane, and change comes from all directions. When we think about a pandemic, we see the effects extending far beyond any actual illness. As we saw with the arrival of Europeans in the Americas, it was not only the diseases that created the catastrophe that followed, but also the breakdown of other systems, such as agricultural systems and social support systems, that exacerbated the effects of the diseases. People take advantage of crises for their own purposes, too, as Naomi Klein reminds us, and that can shape the nature of a catastrophe.[7] We are familiar with the stories of Europeans sending blankets they knew to be infected with smallpox or some other disease to Native American populations that they wanted to damage. In the current crisis, we had political actors such as Stephen Miller withholding help from places that support the opposing political party. While outbreak narratives in the movies might feature an incredibly deadly pathogen, extreme deadliness is not necessary for severe disruptions. We saw this during the Covid-19 pandemic.

While the first two decades of the twenty-first century have seen continuous warfare post-9/11, constant conflicts are nothing new. As of 2021, the United States has been at war during more than 90 percent of its history. For the last seventy-five years, however, the stakes have been raised. Nuclear weapons create the potential for truly world-ending conflict. The Doomsday Clock moves closer to midnight, indicating that the danger of a nuclear conflagration is higher than it has ever been. The advancement of the Doomsday Clock is a reaction to particular political trends. In the past few years, for instance, the rise of authoritarian figures on the world stage, and a concomitant increase in bellicose relationships with other groups, increases the chances for nuclear conflict. Some right-wing political thinkers believe there could actually be a useful nuclear exchange, which would achieve specific geopolitical goals. This willingness to consider the use of nuclear weapons is a departure from attitudes of the recent past and is reflected in the Doomsday Clock calculus.

The likelihood, then, that our current political situation might result in some sort of conflict or even nuclear war can be partially assessed by looking for parallels elsewhere in history. The most concerning to me is the rise of authoritarianism and fascism around the globe. This is especially worrisome in the United States, in Europe, and for other nuclear powers. We can see parallels to what happened in the 1930s with the rise of fascism in Europe, with the type of grievances and reactions that we see on the political stage. That resulted in World War II. Will our current trajectory result in World War III? I doubt that it will, in terms of a global conflict analogous to World War I or World War II. I suspect it would take the form of the kind of conflicts we have been seeing over the last forty years, which are smaller in scale, defined and discussed in different ways but just as potentially worrisome. For me, all of this is eclipsed by the near certain disaster of climate change, already on the horizon. As we see repeatedly in earlier archaeological examples, one event may start the chain reaction, but all of the collapses that we see in history ended up being multicausal. We live in complex systems and we cannot change one part of them without changing another part.

A common sentiment among those who study catastrophes is that there are no natural disasters, only natural phenomena that we convert into disasters via our responses. These natural phenomena happen all the time, and sometimes become natural disasters. They do not typically result in the decline or radical transformation of society. Many examples exist, however, of natural disasters that pushed a region or country down a much more extensive path of change. Hurricane Mitch, the massive 1998 hurricane that killed more than seven thousand people in Honduras, set the stage for a series of problems that fundamentally changed the country from its 1998 reality to what we see today, by exposing weaknesses and opportunities for exploitation that powerful people acted upon.[8] This is a good example of the type of "disaster capitalism" that Naomi Klein discusses.[9] It does not have to be that way, of course. A colleague of mine in Kentucky, Fatima Espinoza, has identified ways in which people were able to improve the systems in place after the disruption of the hurricane, sometimes repurposing technology to replace collapsed systems that did not work well for most people in the first place.[10]

UNLIKELY SCENARIOS

To me, climate change, conflict, economic inequality, and the rise of authoritarian regimes are the most probable contributors to the next apocalypse. There are, however, a number of other possibilities that seem unlikely, or for which there are no ways to calculate their likelihood. These include some of the scenarios we create in our fictional narratives, like the arrival of extraterrestrial life forms. I can say very little about this possibility, except that our limited vision of what this encounter would be like will almost certainly fail to capture the reality. If we look at the Fermi paradox, there is very little evidence that other advanced civilizations exist despite a high likelihood that they do. On the other hand, astronomer Avi Loeb suggests that we have already been visited, and that we have observed alien technology in the interstellar object named 'Oumuamua. That cylindrical object, which was located within our solar system, in Loeb's assessment, seemed to accelerate when it should not have, was highly reflective, and had other unusual properties that made it unlikely to be naturally occurring.

I have no educated guess at the likelihood of any kind of alien encounter. However, what if it did happen? What would that scenario look like? The closest we can come to a guess might be to look at the introduction of a new species or new population into a different population, one with which they have never before had contact. We can look at invasive species, the lionfish in the Caribbean reefs, or kudzu vines here in Kentucky. We could look at the novel coronavirus pandemic. We can look at the contact of people from the Americas with a population from Europe from whom they had been separated for twenty thousand years or more. In all of those cases, the results are bad. I can imagine unintended consequences from the interaction of two or more species that have never met. We do not know whether this would be more catastrophic to one species than the other.

The other thing I could imagine derailing such an encounter would be the difficulty in communication, in interpreting behavior, and in knowing what might prove catastrophic to the other group or provoke a catastrophic response. We do not have a universal translator, and we do not have any experience with extraterrestrials. I suspect any kind of alien encounter would constitute a catastrophe on many levels for many reasons, regardless of the intent on either side.

An arrival from outer space might not be in the shape of a life form but rather a ball of rock and ice. Comets or meteors hitting the Earth are a favorite of apocalyptic narratives. From the data that we have, the chances of this sort of event happening in any of our lifetimes or in the lifetime of any individual in the future is fantastically small and fails to present more than a localized threat. Chances of a large asteroid hitting the Earth is around 1 in 300,000 in any given year, and maybe 1 in 10,000 in our lifetime, according to the Stardate series from the McDonald Observatory at the University of Texas.[11] An impact event most certainly will happen at some point, but it is unlikely to be something about which we should worry.

If it did happen, however, what it would look like depends on several questions. Perhaps the most obvious is, "How big of a thing is it?" What it is made of, how fast it is traveling, and where it hits the Earth would all change the nature of the disaster. Disaster outcomes run the gamut from the kind of minor damage that we see periodically, like the impact of the Chelyabinsk meteor that hit Russia in 2013 to the Tunguska event, also in Russia, in 1908, to the impact of the Chicxulub meteor 66 million years ago near the Yucatán Peninsula, which doomed the dinosaurs and was nearly a global killer. That is, outcomes could range from almost nothing to disastrous for only some people in certain places to disastrous for everybody.

Another scenario that is less likely to be included in these narratives but certainly is in the public imagination would be a large volcanic eruption, perhaps even a supervolcano like the one under Yellowstone Park. Volcanos erupt all the time, and some have devastating localized impact, but global impacts that result in catastrophic events are very small. The United States Geological Survey (USGS) puts the odds of the Yellowstone supervolcano erupting at around 1 in 730,000 in any given year. Volcanos that trigger global catastrophes, like the 1815 eruption of the Tambora volcano in Indonesia; Krakatoa in 1883, also in Indonesia; or the Okmok volcano in Alaska that might have contributed to the fall of the Roman republic in 43 BCE, are less than once-in-a-lifetime events. From our historical examples, we can look at localized disasters and the larger impact they had. Volcanic eruptions like the Ilopango volcano in El Salvador in the fifth century CE changed the course of history in Mesoamerica, changing relationships between groups and possibly leading to the downfall of Teotihuacan, a civilization based near present-day Mexico City.

Random and very unlikely (but eventually inevitable) things like meteors striking the earth and the eruption of a supervolcano are all possibilities that could occur at any moment. Data suggest that such events are unlikely within our lifetimes, however, or for even generations to come. These cannot be discounted completely, but they seem implausible, unpredictable, and remote, especially when compared to the challenges we face now or will see on the horizon.

What the next apocalypse will be like depends, of course, on many factors, but there are a few things about the way it will likely look that I can reasonably surmise based on what has happened in the past. I see a clear contrast between historical catastrophes and the way in which we imagine future catastrophes. Apocalyptic events in popular narratives are sudden. But remember, as I have repeatedly noted, the archaeological record always suggests that historic collapses took decades or centuries. I imagine we will find ourselves in a long-term predicament. It is not a sort of "grab your bug-out bag and rough it for a couple of weeks" type of deal, but rather something along the lines of dealing with months of food insecurity, followed by a lifetime of resetting basic systems, such as agriculture.

Individual skills and efforts will not be sufficient in the next apocalypse, and we will need to work as a community. I surmise that a great number of people would be affected but would still be around and in need of help. I imagine that any large-scale apocalyptic event would involve billions of survivors, not the handful of survivors that we see in many of the apocalyptic narratives. That large number of survivors means that collective action and cooperation is necessary, just as it is now in any community.

The most common criticism of my initial op-ed from 2019 about the next apocalypse was that I somehow did not understand the seriousness of the impending disaster if I assumed billions of people would survive.[12] That criticism has little merit, given that we could suffer a monumental catastrophe and still have more than a billion people on the planet. With nearly 8 billion people alive today, a catastrophe would have to wipe out 85 percent or so of the population in order to get us under 1 billion people. Only in the very worst historical catastrophes have 85 percent of a population died, and then only in a relatively small part of the world. Nothing like that has happened on a global scale in the last 70,000 years—a massive eruption of a supervolcano in Indonesia and subsequent global cooling that caused the global human population to drop to fewer than 20,000 individuals. The most recent example of this sort of catastrophic demographic collapse happened in a more limited geographical area in the Americas with the arrival of Europeans.

My point, however, was not about the actual number of survivors. I meant to emphasize that we will be living in groups, not tiny enclaves of survivors. The groups will be much larger than our fantasies account for. We will not be a small family unit, for instance, able to hunt for our survival. We will be part of a community, with all of the advantages and challenges that presents.

The other thing that my critics failed to realize is that we are going to identify and label an event as apocalyptic far before we get to the point where 85 percent of the people have died. Let us take the example of climate change, with rising temperatures that threaten the world's agricultural system. Great famines will result. It could do us all in, eventually. Are we going to wait until 6.7 billion people have died before we start acting as if we are living through an apocalyptic event? Will we wait till we get to that point to respond to the crisis? Of course not. Even given our current inaction in the face of compelling data, we would reach a point where even the most reluctant were spurred to action. We would begin reacting to the apocalyptic event long before the catastrophe reached its zenith, when there were in fact billions of survivors, even if fewer might be left at the very end. So positing billions of survivors is not naïve. It is inevitable.

In our own lived experience, we do not interact with 1 billion people in any meaningful way. It might be more useful to think about postapocalyptic life in a way that reflects what it would look like as we go through it. If you live in a very low-density rural area where you already have a limited group of people with whom you are in regular contact, the change in your daily experience might be radically different from someone who lives in a large city, where interacting with thousands of people is common. In fact, we often participate in systems that involve millions of people. As we see from the Classic Maya collapse, the Roman Empire, or in the severely impacted North American groups, communities survive. People continued to live in communities, much bigger than the small, clannish group of survivors we see in many of the apocalyptic narratives, and we will, too.

This matters because it dictates how we should respond to a crisis. If I'm the only person on the planet, or if it's me and the four other members of my immediate family, we have a whole set of different problems and goals and needs than if I'm living in a community of one thousand people, or ten thousand people. A small group, with some knowledge, can survive in almost any part of the world by hunting and gathering. Humans did this for most of their history, usually in groups of fewer than fifty. In the larger, more densely populated, world that exists now and will exist even after a catastrophic event, we have much more complex systems that are going to need to be rebuilt at large scale. The most obvious of these is the agricultural system that we use to feed ourselves. Our numbers will not drop anywhere near the small number of people alive the last time all humans were hunters and gatherers (less than 10 million). By the time we had a global population of 200 million, almost everybody practiced agriculture.

How many people survive and how they might be distributed across the landscape will determine the skills and the behaviors that we will need in order to make it through. Examining any of the past catastrophes or collapses suggests that we will have to work together as a community to solve problems. Striking out on our own will not work.

One way to think about what the next apocalypse will look like on the ground is to think about the recovery. For many of the plausible scenarios, if we generalize, a somewhat similar chain of events will take place: Some proximate cause or causes will precipitate an event. This will mature unnoticed, and eventually (it may seem sudden, as the prelude had gone undetected) the complex systems we have created will fail. Governments might fall, or an established social order might disintegrate. Trade routes, economic systems, and agricultural systems might no longer be viable. The specific causes, the particular history, and the particular systems at work at the time and place will determine the details, which will always be unique. History suggests what the recovery might look like.

Initially, there is the immediate aftermath to a dramatic and widespread transformation. This is the phase where the kinds of survival skills I teach might come into play. People engaged in critical activities such as farming or maybe public safety will continue working. As Kara Cooney elaborated on when speaking about the fall of the Western Roman Empire, people might lose their patrons (the ones for whom they work, or with whom they have contracts) and will have to find new patrons or create a similar relationship with the people around them. People with less immediately critical jobs, or whose employers have disappeared, will have to switch gears more radically.

If we look at recovery from catastrophes like hurricanes, we can develop a timeline. There, we have days or weeks of dealing with the loss of basic necessities. After a relatively short time (a matter of weeks), our efforts switch from getting basic needs met to re-creating the elements of society that are critical and important. This rebuilding can take years or decades, and the resulting reconstruction might bear little resemblance to the original.

CHAPTER 8

WHO SURVIVES AND WHY

I held my breath as long as I could, eyes closed, worried about my grip on the wrench. I had already dropped it on the bridge of my nose once. I lay on my back in eighteen inches of mud and water, trying to remove a broken leaf spring from my old Land Cruiser. I sat up a little to breathe. I had to wipe the mud away from my mouth and nose every time I took a breath, which annoyed me, as I could not keep my hands on the bolt I was loosening. With the spring broken, the weight of the car came down on the front tire. The jeep was not moving anywhere, certainly not the quarter mile to get out of this swampy field. I had to change it there. I should have put a snorkel in my toolbox.

I was in Honduras doing archaeological fieldwork for my doctorate. We crossed this field every day. The roads in the area were muddy and rough. We got stuck a lot, and it was hard on the vehicles. Several of the Pech worked with me, excavating archaeological sites all over the valley. There were probably five or six other villages and small towns in this valley, but none of the others were Pech villages. All of the others were on the main road, and getting to them was much easier.

At that time, the Pech village where I lived had no electricity and no water system. None of the other villages had electricity, either, but they all had running water. Not every house had a spigot, but there would be one nearby. Where I lived, we walked to the creek that ran along the edge of the village to get water and to bathe. Every day after work we would walk down the hill, get washed off in the shallow water, then go upstream and fill up six-gallon plastic jerry cans with fifty pounds of water to carry back up the hill to our house.

The water was heavy. The creek was cold. The inconvenience that mattered most, however, was the road in and out. Even with a capable four-wheel-drive vehicle like my old Land Cruiser, hitting the wrong line through the tricky parts meant an hour of digging out. That added up. Also, I failed to calculate into the equation the risk of drowning in a mud hole while fixing the truck.

Despite all that, I always lived in that particular village anytime I worked out there. We became friends, of course, so that was the real reason I came back. But the first time, I had a choice. I'd decided to ask the Pech if I could stay in their village because they were probably descendants of the folks who built the archaeological sites I studied. It seemed like the right thing to do, and issues like water did not really bother me. The time it cost to get in and out of the village meant less research, though, and that was the only real concern.

At first, before I learned the road and bought better tires, we could get stuck every day. I had gotten a couple of grants to do my research, but budgets were limited, and every little bit counted. My crew got paid no matter what we were doing, whether it was archaeology or digging out the truck. I never doubted or regretted my choice, though. In fact, it was the best decision I could have made. I owe so much to the Pech, for helping me negotiate a relatively new place to interpreting the archaeological sites we documented. They knew about every single large archaeological site out in the rainforest, and were humble and funny and gentle, for the most part.

In Honduras, I was not in a survival situation, nor weathering a catastrophe. I loved living in that village, from the pace of life to the way we passed the time. It was there that I learned many of the skills that I now teach in my survival skills courses. The lesson that proved the most valuable, however, and the one that will be indispensable in any of the plausible future apocalyptic scenarios, is the importance of social skills and connections. I learned as much as I did from the Pech in large part because we were friends, we lived together, and we shared a life for a while.

My wife grew up in a small village in Honduras without electricity or running water. When we watch postapocalyptic movies together, she comments on what people think constitutes a disaster (losing the power grid, for instance). How people react to sudden change also puzzles her. In her early life, changes were frequent, the unexpected happened all the time, and material possessions were temporary at best. Planning for the future looked very different for her as a youth than for me as a kid in Kentucky. What constitutes a disaster for her may be very different from my definition, even after all this time together in both of our native places.

For my wife, change is inevitable, and not cause for melancholy or regret. As I discuss this book with her, she rejects the premise that radical change represents some sort of failure or tragedy. For her, societal collapse should be reimagined as society reborn. This is not to minimize the catastrophes and tragedies born of such an event, but to understand that when it happens, it is not new: it has been happening throughout human history. An "apocalypse" is not a mystery. Every day we see small "apocalypses," and we certainly have archaeological and historical records of constant change at every level.

While change always happens, not everybody makes it through. In this chapter, I look at archaeological, historical, and modern accounts of disasters, investigating who survives and thrives, who does not, and what situations and skills contribute to the outcome.

Archaeologists can rarely identify individuals from the past or even get down to the scale of the individual. We deal with groups. We can see the results of individual action sometimes, and of course we see the monuments or iconography that powerful elites or leaders arrange for themselves. We can make some educated guesses as to the status of a person who occupied a particular place at a particular moment in time by looking at the things they left behind, but putting together any kind of life story is incredibly challenging. If we want to look at who survives catastrophic events, we need to first examine some contemporary and historic examples for which we have different sorts of data and documentation.

There are going to be some physical qualities of an individual that might give them an advantage. Overall fitness could be an advantage, including how strong you are, how much endurance you have, and your degree of mobility. In my experience, great improvement in fitness comes quickly, though, so the difference between a very fit individual and an average person might narrow greatly in the first few weeks after an apocalyptic event. When I moved from Chicago to rural Honduras, my daily activities changed from life in a city to an intensely physical, rural life where walking for hours, carrying a load, and working with a shovel or machete constituted a normal day. The first days seemed unbearably difficult. Two weeks into it and I no longer thought about it much. Being in shape will be most important in the first few days or weeks of a catastrophe, but most of us will become accustomed to new physical demands in short order. The social attributes you have, I imagine, will be more important than your physical attributes.

When we think about things that can affect your likelihood of survival in the face of any kind of challenge, I immediately think of socioeconomic status, nationality, race, gender, and sexuality. Some of these variables, like socioeconomic status, might affect access to things that make survival possible or easier, but all of these factors can affect how you're perceived and treated, and that certainly can affect your outcomes.

Let's take a closer look at socioeconomic status. Wealthier people or privileged people face fewer hurdles to survival. This could mean having access to things like medical care or technology, but privilege can also affect whether you would be accepted as part of a particular group or social class. I know, for instance, the reaction that some of my rural relatives would elicit in certain urban, academic settings because their aesthetics or accent would mark them as outsiders. I saw this all the time as a child. As a college student, a professor advised me to lose my Southern accent, because it would hold me back. I think that is horrible advice, but it is true that things like accents shape people's expectations. I experienced this firsthand on several occasions, where my Kentucky dialect was singled out. One instance stands out as particularly egregious. I was at a party in graduate school with a friend of mine. I did not know most of the other people at the party. We spent the evening talking and joking and getting to know one another. Finally, after we established a degree of familiarity and confidence, one of the graduate students, a woman from the northeastern United States, said to me, "It's so strange to hear funny and clever things said with that accent." Nobody within earshot reacted, perhaps not realizing the insult buried in that backhanded compliment. I replied that it was not unusual at all to hear comments like that in *her* accent. That experience was not unique or particularly significant, and certainly did not have any meaningful effect on the trajectory of my life. In another situation, however, where people are deciding if you belong or not, it could be very significant.

We can see these differences in action if we look at the Covid-19 pandemic in the United States, which disproportionately affected African American and Hispanic populations. This was not due to some physical or genetic difference or because of variable reactions to the disease. There were efforts to blame it on these things, as there always are. We know, however, that our racial categories are cultural, not biological, and that the biological differences that exist between groups is less than the variation that exists within a group. Therefore there's no way to pin what we see happening on biological differences.

The reason for this disproportionate negative impact on Black and Brown populations in the United States seems obvious, but it is not simple. Poverty and the related stress of being poor affects health, creating distinct conditions among populations of lower socioeconomic status. Lifelong trauma takes a toll. Add to this a general lack of access to medical care, or underlying conditions that have many other causes, and you set the stage for very different outcomes among different populations.

We see the difference when we look at the reaction to Hurricane Maria in Puerto Rico or recent earthquakes in Haiti. How a group is perceived will influence the response mobilized to help them, by other countries and even their own government, as with Hurricane Katrina. These responses have racial overtones, of course. The Trump administration's response to Hurricane Maria in Puerto Rico was not only horrific, it was also based on the same racism that fueled Trump's anti-immigration sentiment. In fact, the idea of what an American should look like is so strong that on repeated occasions, government officials implied that Puerto Ricans were not U.S. citizens.* These responses demonstrate the way in which socioeconomic status and race (or, in other cases, caste or religion) can position people in easier or more difficult positions.

* While Puerto Rico is not a state, Puerto Ricans are citizens, of course, and have been since 1917.

We see some hints as to what qualities facilitate survival by looking at archaeological examples. In a conversation just before the beginning of the pandemic, archaeologist Chris Pool talked with me about his work on the decline of ancient civilizations in Mexico. He found that areas with evidence of flexibility in terms of how they organized themselves were able to survive the changes. For instance, areas with greater flexibility and adaptability in the scale and location of centralized leadership were able to continue longer than in areas where these variables appeared more rigid.[1]

The archaeological record also supports the idea that when we talk about a collapse, we are usually talking about the abandonment of some of the larger urban settlements or cities. What we see is people moving into less densely occupied zones. That seems to go along to some degree with the opinion in the prepper community that cities are the wrong place to be in a disaster. During the pandemic, for instance, cities were harder hit and the initial shortages of essentials like toilet paper and hand sanitizer were felt more acutely there.

In addition to the importance of flexibility and the tendency for people to move from urban to rural areas, archaeological examples suggest that the changes are not felt equally throughout society. Archaeologist Gwynn Henderson captured this during a conversation we had about previous collapses. "The bosses come and go. The top of the pyramid is chopped off, but the regular people are still there, doing their thing—eating, living, and reproducing," she noted. "Isn't it ironic that history is all about the people who are expendable? The heads come and go, but the people are always there." Her sentiments reflect Kara Cooney's comments that when things collapse, people lose their patrons, and they have to find new ones, but otherwise they might have a lot of continuity in their daily life.

The details that we see in the past and what we are likely to see in the present or future may be very different, because the systems in which we live are so distinct. We saw in the Maya area, for instance, that the drought at the end of the Preclassic had a very different result from the drought at the end of the Classic Period. That difference has to do with the nature of the systems in which people were living. There was a much more complex social and political system in place in 900 CE than 800 years earlier. When we look at the present day, we have to understand that we are living in a globalized world with systems that are much different from what came before. A few key differences come to mind.

The first difference has to do with property. Ideas surrounding ownership of land change over space and time, and were undoubtedly different from the sense of ownership that we have today. When I think about bugging out, grabbing a backpack, and heading up to the hills in the event of a disaster, the first thing that comes to mind for me is the reaction of the property owner. Will they welcome people in need? Will you be shot for trespassing? The rural/urban divide in the United States has a racial component, and that could matter. I anticipate that moving out of the city into the countryside will be very different in the United States than it would've been in the outskirts of Tikal, Guatemala, some twelve hundred years ago.

Another big difference is our agricultural system. When we think about making it through some sort of apocalyptic event, the first of our longer-term concerns is food. Our current agricultural system is radically different from any system in the past. First, we have a degree of monoculture or single-crop farming never seen before. Much of this relates to farm subsidies that were set up to encourage this kind of agriculture for export. When I go for a drive in rural Kentucky, I see corn and soybeans. Years ago, I would have seen much more tobacco. When I was in graduate school in Chicago, rural Illinois was the same. There have always been principal or "staple" crops that dominated any agricultural system. In Illinois one thousand years ago, you would have seen predominantly corn. There was also gardening and small-scale agriculture too. To some degree we are seeing a resurgence of the popularity of gardening during the pandemic. By and large, however, we have a much different way of growing food than we see in history, and this will certainly influence how we react to a catastrophe.

The next big difference involves who practices agriculture. Industrial agriculture, and the decline of the family farm, has changed who participates in agriculture, and to what extent. This shift has been greater in some places than others, but it has been a worldwide trend for decades. There are a couple of important implications here. One involves who owns the land and who would have access to it in the event of an emergency. In case of a collapse, the industrial interests could lose control, and access to the land might be wide-open. It could be just the opposite, however, where powerful entities shut down access to their land and moving from the city to the countryside would have limited benefits. There'd be nowhere to stay, and no land to cultivate.

Most people's general lack of agricultural expertise would be especially important in an emergency. It is possible to learn how to do many basic tasks in any endeavor in a short time. I can learn to weld in a short amount of time, for instance, but to a limited degree of competency. Likewise, I can research and learn how to plant and grow crops. I know, however, that there is a world of difference between what I might be able to produce with that kind knowledge and what somebody who spent their entire life doing agriculture might be able to produce. It reminds me of the scuba diving that I do as part of my archaeological research. Most of the time, scuba diving consists of breathing in, breathing out, and repeating that process. Simple. You try to keep your heart rate down and use as little energy as possible, and the gear you have on controls your buoyancy, so you float, kick a little, and breathe. It could not be simpler, most of the time. However, when something goes wrong, everything changes. Your training and knowledge kick in; you have to fight the panic, and experience matters. It could be the difference between life and death. Likewise, agriculture in the best of situations might be easy, conceptually, even if it requires a lot of work. Knowing how to manage the hard times, however, only comes with experience.

In developing countries, up to 80 percent of people are directly involved in agricultural production. In developed countries, it is below 5 percent.[2] In the United States, for instance, it is around 1.5 percent of the people. This means that most people, me included, have no experience growing food on a scale that would be sufficient for survival. I do not have experience harvesting or storing agricultural products, or selecting seeds for the next crop, or knowing which landforms are best for certain crops. Contrast that with an experienced farmer, and it is easy to imagine how different the outcome might be. In archaeological examples, preindustrial populations had a much higher percentage of people engaged directly in agriculture for most of their lives, with a level of competence we rarely see in contemporary times.

The last big difference, and perhaps the most obvious, is the global population. Nearly 8 billion people live on Earth today. By comparison, there were fewer than 500 million people in the world in the sixteenth century. The Earth had less than one-fifteenth of its current population, or about 7 percent, during any of the archaeological examples I explored. I imagine that is going to make a significant difference. One of the criticisms of certain parts of the prepper community and the whole bug-out mentality involves the degree to which people could actually find food in a non-agricultural situation. We know that most humans have lived as agriculturalists for the last three thousand years, upward of twelve thousand years in some places. This transition from hunting and gathering or foraging to agriculture was not one that people made easily. Marshall Sahlins, one of my professors at the University of Chicago, wrote about the lifestyle of hunters and gatherers versus agriculturalists.[3] He noted that modern hunters and gatherers, living in harsh areas not suitable for agriculture, like deserts or the Arctic, typically work fewer than fifteen hours a week to satisfy subsistence needs. Criticism of his account challenged those numbers and the concepts employed, such as affluence.[4] Contrast this estimate, even if you double it, with the amount of time that farmers dedicate to agricultural activities a week. In terms of time and effort, I understand why people might resist a transition to agriculture. There are offsets, including the security of growing more than you need, of storage, and perhaps the security of living in one place. Hunters and gatherers were also growing plants to some degree. After all, they made their living gathering plants. People did not *discover* how to do agriculture. They already knew how plants worked. They started using agriculture when they had to supplement their hunting and gathering, because population density increased beyond what was feasible to support through foraging. Intensive agriculture produced far more food than what can be procured from hunting and gathering. As I mentioned, by three thousand years ago, most people had adopted agriculture, and the population of the world was less than 10 percent of what it is today.

Almost all people before agriculture were nomadic or seminomadic; they did not live in permanent villages. There were two exceptions to this. One was on the coast of Peru, where there are massive amounts of seafood available because of the upwelling Humboldt Current bringing nutrients from deeper in the ocean to the surface, resulting in enormous quantities of sardines and other small fish. These were caught using nets. In fact, some of the first agriculture included cotton, which was used to make nets to get food from the ocean.

The other place where people were able to live year-round in one location before the advent of agriculture was here in Kentucky, along the Green River. People used riverine resources, such as mussels, enabling them to stay in one place and resulting in great shell middens—or piles of discarded shells—from this activity. However, outside of those examples in Peru and

Kentucky, even at low population densities, hunting and gathering was not enough to sustain people in one place year-round. When we look at who survives and why, we need to take into account the realities of our current agricultural system and how food is distributed.

So I have discussed the personal physical attributes, the social systems, and the skills that one brings to the table, but I have not yet mentioned the one thing that might be the most important: the will to survive. We know in more acute, short-term survival situations that the importance of the will to live cannot be overestimated. Your survival instincts do not necessarily kick in automatically. Just as it might take intense effort to muster the will and energy to make it through a short-term survival situation, long-term survival can be even more exhausting. Summoning the will to survive would be one of the most important factors in terms of who survives and thrives in an apocalypse. People with a reason to keep going will have a much better chance of persevering. In my case, and I suspect in many cases, that reason would be family.

Most of us have an understanding of the role of strong family connections and friendships for both physical and mental health. That will certainly be true in an apocalyptic situation. A strong network of family and friends provides resources, knowledge, and safety in numbers not available on your own. One thing that always resonated with me about the book *Lucifer's Hammer* is the way in which the street gang leaders retain their leadership status, and followers, in the aftermath of a comet strike that nearly destroys the Earth. Similar situations are found in any number of postapocalyptic or dystopian narratives, where street smarts enable you to gather a group of followers to work for you, from Negan, the leader of a group of survivors called the "Saviors" in *The Walking Dead*, to Carnegie, the de facto ruler of a town in *The Book of Eli*. The idea of social competence resulting in access to a group of followers is an old trope. Charismatic leadership skills cannot hurt. In terms of preparing for a survival situation, those skills may not be particularly useful, however. Also, charismatic or political leadership may or may not be a skill you have or can develop. Most of us are not charismatic leaders, capable of persuading a group of people to follow us.

The other and bigger problem occurs when everybody wants to be a leader. We need leaders, but we need followers to do the work once a decision has been made. I see this dynamic in the world of nonprofits, where everyone wants to start a new organization and be a leader. Duplication of effort wastes resources and frustrates those already working on a problem. Sometimes, leaders and organizations emerge because they embody new ideas or tactics. Other times, it is a manifestation of ego and desire. The will to power may be especially damaging when resources are scarce. "*Yo no soy marinero, soy capitán*" (I'm not a sailor, I'm the captain) sings Ritchie Valens in "La Bamba," and that attitude will hurt us in the aftermath of a disaster.

Social groups go beyond leadership and charisma. It will be important to be a good community member—a follower as well as a leader. When my college students work on group projects in class, I see a certain type of student who routinely emerges as a "leader," where the scare quotes mean a self-styled leader. These students assume leadership roles without any consensus among the group that they should be in that position. These are the confident students, the loud students, and the students who have had social success and are used to being noticed. Most of them are members of privileged groups: white people, males, people with higher socioeconomic status, and people who are considered attractive for some reason.* I see people with good ideas shut

* While there is some biological imperative to certain elements of what we find attractive, we have to remember that the powerful are defining attractiveness to fit whatever qualities they themselves possess, so attractiveness and power are not completely separate.

down or struggling to be heard. Sometimes, a student emerges as a leader, an effective leader, only after a lot of time and energy has been wasted with the blustery self-appointed leader. When the stakes are higher, and when survival might be on the line, a shallow façade of leadership will be less acceptable than ever. Also, as a member of a community, as a follower, one of our responsibilities will be to make sure that the people with good ideas have an audience, and that leadership is not hijacked by the overconfident ones who speak up quickly, but without expertise or knowledge. A cartoon I saw in *The New Yorker* captures this. It shows a man interrupting a woman, saying, "Let me interrupt your expertise with my confidence."[5] We cannot do that, or let it happen.

In considering the importance of competent leaders, we need to maximize the pool of candidates. That is, we need to make sure that our definition of leadership does not exclude people. Sometimes we define a thing based on our experience. If that experience comes from within a system in which certain people are excluded or marginalized, our definitions will reflect this and only further the inequities of the status quo. In an earlier chapter I discussed the ways in which some people in the prepper community would ask about the skills you bring to the table in order to admit you to their group. In those examples, prejudice, bigotry, racism, or sexism is embedded in the question. The privileging of some sort of skill serves as a proxy to weed out undesirables. I am reminded of the literacy tests that were once given at polling places in the United States, which were designed explicitly to prevent African Americans from voting, rather than as a test of literacy.

Skills are valuable, though. We want people with skills. Some sort of vetting might ensure you have people with useful skills, but it might not be necessary. People learn, and quickly. The utility of somebody to a group lies more in their *potential* than in the skills they already possess. I see no justification for any kind of test of utility because utility should be assumed to be there, undeveloped but not out of reach. Some narratives suggest that resources might be so scarce that we must choose who survives, and we can all imagine some situation where the constraints are such that we might need to choose one person over another. We will not find ourselves in those situations. If we are tempted to exclude people, we must find another route through the disaster. We must include everybody, or we are not in a sustainable pattern. Rather than reject people, we need to allow them to develop into valuable members of the group. People will rise to the occasion.

Adaptability and flexibility will be key to survival. Surviving and thriving after the next apocalypse will be all about community. None of us will be able to go it alone. We will not be able to try to re-create what we lost, and maybe we won't even want to. Things will be different, and people will have seen the collapsed structures and systems for what they are. They will have seen behind the curtain, and they may not want it anymore, so we must be flexible moving forward.

We are going to have to get comfortable with being uncomfortable, and not just in physical ways. We might be colder, hungrier, or sicker than in our life before. Living spaces, sleeping arrangements, and daily activities might be incredibly uncomfortable and unfamiliar, but there will be other things to which we must adjust. We might face new and unfamiliar ideas about race, gender, or sexuality. We may be interacting with people with beliefs we do not share, a worldview we do not understand, and customs that are unfamiliar. Our ideas about personal property or personal space may be challenged. All of these things have changed over time and space, throughout the history of humanity, and an event catastrophic enough to be called an apocalypse might create the need or desire to change these ideas again, only in an accelerated way.

BUSHCRAFT OR SURVIVAL SKILLS

On the day of the coup in Honduras in 2009, I was walking into the rainforest with a group of students. We stopped at the last mountaintop where we could get cell service, and I called a friend in the capital. For a couple of weeks, it had been rumored that on this day, the day that voting was scheduled for a constitutional referendum, the president would be removed by the military. I called my friend to see if the coup had happened. It had.

This is bad, I thought. I had ten college students to worry about and keep safe. I did not know what to expect, but I was certain that ten sets of worried parents were calling the university, asking for information that was not available. We were traveling through an uninhabited area, so I thought we should be all right for several days. I had a satellite phone, but I could not get a good signal due to heavy rains, thick cloud cover, and the narrow valleys we were walking through. I feared that a student would get cut off after a couple of words with a parent, making the whole situation worse. After a couple of days, I was able to contact my wife, who told me that there were violent protests in the capital, and that networks worldwide were broadcasting these scenes twenty-four hours a day. She was able to contact parents, so that at least they knew their kids were all right for the time being.

I had no idea what we would find when we walked out of the forest. I was not even certain we were safe far out in the forest. Whenever I would call home on the sat phone, I would walk twenty minutes or more away from camp, because, not knowing what was going on in the rest of the country, I was afraid that the sat phone signal could become a target for a missile or an air strike. That sounds silly in retrospect, but it didn't feel silly then. For the next week, traversing the rainforest, heading for the coast, I came up with an escape plan if it looked like Honduras was heading into a full-blown civil war. Near the coast, I knew of a hill I could climb to get a cell-phone signal, which was much better than the satellite signal. I would call ahead to the capital to see how the coup had unfolded. If it was grim, we would head south through the swamps to Nicaragua. It would take us another week. The trip would be grueling and incredibly unpleasant, but it seemed infinitely better than leading a group of students into a war zone. I climbed the hill, called ahead, and found that there was not widespread violence at that time, and we did not have to detour.

I had been ready to flee, though. I had even divided essential gear and supplies among the students, just in case. I understand the instinct to gather your people and head for the hills. I was not thinking about community or helping rebuild or anything like that. I just wanted to keep everybody safe. That had to be my first priority. I consider the type of wilderness survival skills that I teach to be marginally useful in most of the apocalyptic situations we are likely to face, but marginally useful is better than completely useless, and in the early days of a catastrophe, they might be critical.

"You were waiting for this, I guess," one of my students said to me near the beginning of the Covid-19 pandemic. She saw people hoarding supplies and figured that this was the sort of crisis I was prepared for. I was not waiting for this, however. In fact, my preparations and the skills I teach were not useful during the pandemic and would only be useful in limited situations. Certainly in the immediate aftermath of a disaster, life might require expedient solutions to fundamental needs, and our convenient ways of doing things like cooking or staying warm and dry may not be available. However, before long, those skills will just be background activities, akin to doing laundry. The real work and skill that's necessary will involve feeding and housing everybody, educating the population, and doing other things a civil society should do, like protecting the vulnerable or equitably distributing resources.

As an analysis of our apocalyptic fantasies demonstrates, we like to imagine that self-reliance and old-fashioned skills will be the key to survival when it all goes wrong. While I do not think the type of bushcraft skills I teach in my survival courses would be the solution to the types of problems that we will face in the next apocalypse, as I said, they won't be useless. The immediate concern will be safety, as it was for me in the days after the coup in Honduras. Even if you are entirely other-focused, you cannot help anyone if you have died of exposure or thirst.

Bushcraft skills, what we often call survival skills or disaster preparedness skills, will be useful in a limited but important way. I use the term "bushcraft skills," although billions of people use many of these skills (such as fire building) daily all over the world. In many possible scenarios, I imagine the utility of these skills will be concentrated in the days immediately following a catastrophe. They might even serve in the long term as a low-tech alternative to something we currently do.

The ability to build fires or construct shelters may be very valuable where dwellings have been destroyed or the weather makes exposure to the elements dangerous. Having confidence in your ability to adapt and overcome adversity and to use the minimal resources on hand to solve essential problems might make a significant difference in your situation after a disaster. These skills could mean the difference between surviving or not, and in any case, these abilities will empower you and increase your options.

Looking at historical and modern examples, we see that communal efforts will be the key to long-term survival. Learning bushcraft skills can be important for the community, not just for you and your immediate group. If you know these skills, you can teach others. While they are very easy skills to acquire, they can be much more difficult if you do not have a teacher. I learned from Jorge Salaverri and Mariano Alcantara in Honduras, from Ray Mears during the filming of our documentary, and from Craig Caudill back here in Kentucky. I did not have to go through much trial and error myself. More important, having those skills allowed me to take on the role of teacher and spread that knowledge among people who might need it. You could find yourself teaching these skills, not just using them. That could be valuable as well.

There are hundreds of books dedicated to bushcraft skills, and most have similar information. My approach to survival training has always been to offer a decision-making paradigm, rather than merely a set of skills. These decisions include whether to stay put or move, and how to provide basic necessities like shelter, water, and food. The particulars of the situation will determine this, of course. The next task is to take stock of useful items or stores, which might also suggest a particular course of action. The availability of clothing, food, and water could radically alter your plan.

In the decision-making paradigm that I teach, the first decision is always whether to stay put or move. In most cases, staying put will be the correct response. In a short-term survival situation, especially one in which you are lost, the decision comes down to the possibility that somebody will come looking for you. If nobody knows where you are, they will not come looking. If there is no search-and-rescue infrastructure, you may need to move. If your current location is unsafe for some reason (such as the possibility of flash flood or wildfire), you may need to move. In longer-term situations like I am envisioning for the next apocalypse, leaving (or "bugging out") seems to make little sense. The sorts of things that need to be re-created will require collective action in a community setting, and to me this suggests that sticking around and solving problems will be the necessary course of action.

In most emergencies, like a plane crash or getting lost in the woods, your best bet is to stay in one place, put out signals to help people locate you, and wait for help. Sometimes you need to move. In those instances it might be helpful, or even critical, to be able to tell directions from celestial clues or information in the landscape. In terms of celestial information, during the daytime you might be limited to looking at the sun. Almost everyone knows the sun rises in the east and sets in the west and traverses the sky in an arc that does not actually pass directly overhead but rather in the half of the sky toward the equator. If you don't know what time it is, you can calculate the time of day by the position of the sun. Imagine an analog watch. Point the hour hand to the sun, and then find the point halfway between the hour hand and where that hour hand would be at noon (backward or forward to the 12), and that is south if you are in the Northern Hemisphere, and north if you are in the Southern Hemisphere. At night, you have more options. Polaris, the North Star, is faint and often hard to see. Most of us are familiar with the Big Dipper, on the other hand. The cup of the dipper is more or less north, hence the "follow the drinking gourd" story that enslaved people recounted to help guide them to freedom north of the Ohio River, the boundary between free states and slave states.[6] If the moon is visible and not full or new, you can draw an imaginary line between the two points of a crescent moon, and then extend that line down to the horizon. Where it meets the horizon (which may be some distance from the moon) is south if you are in the Northern Hemisphere and north if you are in the Southern Hemisphere. You might find clues in the environment too. The sun shines from the south if you are in the north, and plants that like less direct sun, like moss, will be more common on slopes that face the other way, or on the side of plants away from the sun. Prevailing winds, west to east in most of North America, shape plants as well. Of course, local features such as hills, cliffs, and other plants could affect the way in which plants grow, so you may need to observe a number of plants and take some sort of average. Armed with these techniques, you can (roughly) determine direction in most cases.

A decision that we would face in an emergency situation is how to keep your body temperature regulated. Your body regulates its own temperature, actually, so the goal would be to keep within the parameters where your body can successfully regulate its temperature. Our efforts to stay within those parameters I call shelter-building here, but other considerations include clothing as well. If it is too hot, there is little to do other than find protection from the sun, find the coolest place (a grassy area rather than blacktop, for instance), minimize exertion, and stay hydrated. If we cannot find or create shade, clothing has to do the trick. A hat is essential, and in the absence of one, you may need to rearrange the rest of your clothing to make one. There is no special trick here. Light-colored clothing reflects more heat, and thick clothing may hold too much heat in. Clothing can be used in various ways. It can be worn as intended or used to rig up a shade that will protect you. To create a shelter in a hot environment, the key is to provide shade but not block any breeze that might aid in cooling down. Getting wet and letting the water evaporate will cool you down also.

In cold weather, we have the opposite problem. Trapping our own body heat and sheltering from wind and rain will be the challenge. Clothing is our first line of defense, and having access to appropriate clothing could be critical. We often hear that some huge percentage of body heat is lost through our heads. The numbers often cited are not accurate (45 percent of body heat is not lost through the head, as commonly stated), and you will not lose heat faster from your head than from many other parts of your body. However, given the standards of dress for many of us, our head might be exposed while the rest of our body is covered. If possible, cover that last remaining exposed area. However, if you have no way to do that, take comfort in the fact that you will not lose an inordinate amount of heat; you are not doomed.

You must get dry and stay that way, however. Thermal conductivity in water is around twenty-four times that in air, so being wet will drastically accelerate heat loss. Some clothing retains its insulating qualities when wet, like synthetics and wool, but everything is better dry. You must also get out of the wind. Like water, wind accelerates heat loss, as we know from wind-chill charts. The cold you feel because of wind chill is not merely some illusion that it is colder: the wind will rob your body of heat quickly.

If you do not have proper clothing, or a shelter in cold weather, there is no magic formula. A couple of expedient solutions are plausible using materials that are widely available. First, building a fire can provide the heat that proper clothing and shelter would normally afford. That is our next topic. The other thing is to find some way to improvise warm clothing or a shelter. The first thing I would look for in the absence of regular clothing or shelter would be a plastic garbage bag or sheet of plastic that can be worn as a shirt. In the case of a garbage bag, punch holes for your head and arms and slip it on like a shirt. Most any garbage bag will reflect back a large percentage of your body heat, similar to Mylar emergency blankets. They also protect you from rain and wind. Lying down on a thin plastic sheet without any kind of insulation between you and the ground drastically decreases its effectiveness, but standing or sitting up will yield a real, noticeable difference. Now, this is clearly not a long-term solution, but like other bushcraft skills, it could get you through the first few days. Body heat from other people might help, but all evidence I found suggests the effect is minimal. Cows cluster together in the field to limit exposure to wind and to share body heat, and we can do that, too, even if the principal effect is the comfort of being with other people. Sharing shelters or sleeping bags means you need fewer of them, so this may be where the real benefit lies. Actions like putting naked people together under blankets does not help, and it has the potential to create other types of trauma in the aftermath of a disaster. In these situations, people might feel fragile or powerless, and we must be cognizant of that. I would be very careful during such times to avoid anything that exacerbates existing inequalities, power inequities, or patriarchal injustices. We need to pay attention to the situations in which we put ourselves and others in the aftermath of a disaster.

The other essential bushcraft skill related to maintaining body temperature is fire starting. There are infinite ways to make fire, but I want to focus on three things: what a fire needs, how to construct a fire, and how to light it. First, we need to understand that fire needs fuel, oxygen, and heat to begin and continue. Fire is a chemical reaction, so think about it in terms of how to get that reaction started, and how to keep it going. The oxygen should be available from the ambient air. If it is not, you have bigger problems than body temperature. For our purposes, we need to make sure that the fire is built in such a way as to allow plenty of air in to feed the reaction. The heat that is needed does not refer to ambient heat from the surrounding area, but the intense heat of a spark or flame to start or spread the chemical reaction, the combustion.

To get a fire started, you will need some sort of tinder. Tinder is a term for something that catches fire very easily, often from just a spark. Some kinds of dried leaves or pine needles can function as tinder, but experimentation might be required to see what works well. Very fine wood shavings or sawdust can work. There are volumes written about what works as tinder and how to find it. In my experience, finding tinder in a survival situation can be frustrating and difficult. Things that look like they should easily catch fire, don't. To get around this, I usually carry cotton balls in my kit. Pull them apart a bit to create a loose, fuzzy area (the same can be done with anything cotton) and they start very easily from just a spark. If you are around a car or airplane, fuel or oil might be used to assist in lighting a fire. In a kitchen, cooking oil, or even greasy snacks like chips, can be lit with a match. Hand sanitizer with a high alcohol content is flammable and can catch fire with a spark.

We have all seen people start fires through friction, by rubbing sticks together. Starting a fire from friction is difficult and requires a great deal of practice to become proficient. It is tedious in the best of conditions. I joke with my students that friction methods of starting fires work best in conditions where you least need a fire—hot and dry. You do not want to be caught depending on these methods in cold and rainy conditions, where you really might need a fire for warmth. If you practice this continuously, you may become very proficient at it. Otherwise, it is very hard to do. This should really be a last resort and is considered a long shot unless you practice often. I typically use a disposable lighter or a ferro rod that creates a shower of sparks to start a fire.

When you are ready to start the fire, make sure you have gathered all of your materials in one place first. You need tinder to start the fire, kindling to build it up, and then larger fuel to keep it going. You do not want to struggle to get a fire started and then have to look around for fuel while the fire dies. Unless you do this a lot and have learned how hungry a fire can be, you need more fuel than you imagine. Get a big pile of fuel. Once a fire is going and produces coals, you can leave it unattended for a longer period in order to gather more fuel. In building your fire, you might need to prepare a floor below it, to keep out the cold and wet. This could be a platform of sticks, or really anything dry. You must keep your stack of fuel loose enough to allow plenty of air in to feed the reaction. For fuel, there are many things that could work, but usually we use wood, paper, cardboard, or something similar. Most of us have a sense of what burns. Paper and cardboard are excellent, of course, as is dry wood or wood with lots of flammable sap, like pine. One nice rule of thumb for collecting wood for a fire is that if a limb makes a sharp cracking noise when you break it, like the pop and crack sound a fire might make, it is likely good firewood.

In addition to keeping us warm, fire has psychological benefits. When the sun goes down, a camp with a fire is very different from a camp without one. The light, the noise, and the mesmerizing dance of the flames contribute to our enjoyment of a fire. If we make a fire in our fireplaces at home, it is unlikely for the heat but rather for the aesthetic. Fire-making is easy, with practice, and the result is something that we enjoy in addition to being critical to survival in some cases.

When teaching wilderness skills, I see the connection between the natural world and the cultural. In my survival-skills courses, I usually start with how to build a fire This is always the highlight of the day, the most memorable part. The simple ability to build a fire resonates with people, giving them confidence that they could actually survive an emergency outside. Jack London titled a story "To Build a Fire" for a reason. However, it is also about a connection, or reconnection, with some aspect of the natural world from which people feel estranged. There is some irony in that, as nothing is more "cultural" than building a fire, perhaps. The process of converting the natural to the cultural, however, is energizing, liberating, and often profound.

After protecting yourself from exposure and maintaining your body temperature, the next concern is hydration. Unfortunately, in most situations, there are not many shortcuts for this. Depending on your environment, you might be able to find standing water easily enough. You might be able to distill water from something containing moisture in a solar still, where the sun evaporates the moisture in an item, like vegetation, which condenses on a sheet of plastic placed over or around it. Placing leaves in a plastic bag can produce some water, as the transpiration from the leaves of plants allows water to condense, much like a solar still. In many other situations, plants might signal where water exists. In dried creek beds in an arid environment, plants growing in a low area might indicate subsurface water within easy reach. Some plants will suck water from deep underground, and cutting them will result in accessible water. We did this with banana trees in the

Honduran rainforest in the BBC documentary I made with Ray Mears and Ewan McGregor. All of these techniques require knowledge of the local flora.

After you find or generate some water, you may need to purify it. In a few cases, as with the water that percolates up through the banana tree or water you have distilled from something inside a plastic bag or solar still, you will not need to purify it: the process it just went through has taken care of that. In most other circumstances, however, you will. In many survival situations, the only surefire way to purify water is to boil it. For that, you need a container and fire, typically. Carrying a metal water bottle or a metal cup (like the military canteen and canteen cup system) makes life much easier. Of course, you can possibly find an aluminum can discarded somewhere, or create a vessel out of a piece of aluminum foil. Having a metal container on hand is best.

To boil water, you need to make a fire or have another source of heat. If you cannot make a fire for any reason, there are some other ways that you might purify or improve the potability of the water. Chemical contaminants are not removed with these methods, so finding the cleanest water possible remains important even when you can boil it. You can expose the water to sunlight in a clear container, which allows UV light to kill microbes. This requires several hours and strong sunlight to be completely effective, but anything helps. Some common household chemicals, like liquid bleach, will purify water. Eight drops per gallon, or two drops per quart or liter, will suffice. In the absence of a way to measure out drops, if you can faintly smell chlorine in the water, there is sufficient bleach in there to purify it.

My survival courses typically focus on surviving short-term emergencies, like becoming lost in the woods. I advise people to forget about food for the short term. You will be hungry and weaker, but significant physical and mental declines from hunger take many days to manifest, and eating something poisonous or that might otherwise incapacitate you would be much worse in the short term. For longer-term situations, the best rule of thumb is to eat something that moves (that is to say, animals). Most animals, including insects, birds, and fish, can be eaten if cooked well. Insects might be the easiest to catch. Unless the situation is dire, avoid venomous insects. Again, this is probably a relatively short-term solution, especially for a group of people. Adding plants to the mix greatly increases your possibility of eating something sufficiently toxic to create meaningful negative consequences. Living off the land using plants is a skill you have to develop for a particular place, and it requires a great deal of study and practice. In North America, for instance, there are approximately twenty thousand species of plants. Hundreds of those plants are poisonous, and many more are toxic during at least part of their life cycle. Some are deadly poisonous. There are ways you can test this in an extremely grim situation, but it is not something you want to have to learn in an emergency.

Food contains a lot of water, and if you are not eating much, you may need to drink more. Eating also has a psychological effect. The ability to sit down with something, anything, and eat a little can make an enormous difference in your attitude and outlook. In my courses, I recommend two plants that most people know and that have no toxic lookalikes. These are dandelions and pine needles. Any part of the dandelion can be eaten, raw or cooked, and the few lookalikes that exist are similarly edible. Pine needles can be boiled or steeped to create a tea. Dandelions are nutritious, and contain potassium, calcium, iron, magnesium, and vitamins A and C. Pine-needle tea has little nutritional value outside of vitamin C, but the act of sitting down with a hot cup of relatively pleasant tea has many potential benefits beyond nutrition, including maintaining a good mental state.

Bushcraft or survival skills are relatively straightforward and simple. Even difficult tasks like starting a fire in the rain, or complex tasks like setting traps for animals, are easier than negotiating the social and political world in which we live. This is going to be true in any plausible apocalyptic scenario. Bushcraft skills may be critical to survival for a short time, but eventually these skills will form a background set of talents that everybody possesses. Everybody knows how to operate a cell phone, and most have no problem with complex skills like driving or operating a computer. If I tell a story about an adult in the United States, you will assume that they know how to drive: if they did not, I might mention it. Likewise, very quickly, everybody will be able to start a fire, create a shelter, or perform a formerly unfamiliar skill with perfunctory competence.

What will determine your place in a new world order will be the same things at play now, and those are our position in the structures and systems in which we already live, and the social and political skills that we possess as individuals. Structural and systemic inequalities will remain, or new ones will arise. Sexism, racism, and other ugly tendencies will not magically disappear. Social and political skills, and more immediately, how you treat people, will be most important in ensuring survival in the end.

During the next apocalypse, one of our tasks will be reconnecting with a world we lost. Just as building a fire creates a sense of connection with nature, we will crave that which connects us culturally with our previous selves. Some of the skills necessary for this will be the kinds of things I teach in my survival classes. How to live without electricity, or with less electricity, could require some of that knowledge. The challenge, however, will involve learning to maintain a semblance of our current life—the important parts—in a radically different context. A radical change in the level of technology we use every day seems as if it would be profound, and much of the focus on how to survive the apocalypse involves life skills for a preindustrial world. Based on my experience, however, I do not anticipate that shift to be as significant a challenge as we imagine.

In graduate school, I went from living in an apartment in Chicago to living in that small village in Honduras with no electricity, running water, or even outhouses. Almost as soon as I arrived, I had to wash clothes by hand, bathe in a creek, and start fires for cooking and light. I ate beans and rice for every meal, and spent most of every day walking, digging, and carrying a heavy pack. I adapted quickly, and soon I did not think about these things very often. My life revolved around the people I lived and worked with, who had become my friends. I never once missed electricity or running water, although an outhouse would have been nice. I suspect that adapting to the new day-to-day routine after a society-wide change in technological level will be easier than it seems. The greater challenges will be negotiating the political and social situations, radically changed by migration, food shortages, warfare, and environmental degradation.

As I interviewed various experts in the preparation of this book, I always asked what they thought might be the most important skills in a future apocalyptic situation. Several of them, including all of the archaeologists I talked to, suggested that it would be the ability to recognize competence and the ability to evaluate data and information. Knowing whom to listen to, and what sort of data or information is compelling, is something that we lack even today when we look at topics like climate change or the Covid-19 pandemic. Unsustainable behaviors that exacerbate a bad situation result from listening to the wrong people, distrusting experts, using the wrong criteria to evaluate data, or committing logical fallacies in the evaluation of information.

Conspiracy theories attract many people. Sometimes this attraction emerges because of people's inability to know whom to trust and how to recognize a compelling argument. The psychological makeup of individuals might play a role in why some people become believers of conspiracy theories, while others retain a more critical outlook. Part of the popularity of conspiracy theories relates to how powerful or how marginalized a person is. Being part of a conspiracy theory creates a feeling that you know better than the experts do, or that you are in some kind of privileged group that knows the *real* truth. You become important and special. For academics like me, conspiracy theories don't tend to be attractive. We are already in a privileged group, and we present ourselves as arbiters of truth. Most of us in academia come from groups with one type of privilege or another, reflecting larger systems and structures of power and inequality. We can see this by looking at the racial, gender, and socioeconomic makeup of college professors in the United States, or anywhere else for that matter. Perhaps academia's exclusionary practices encourage those left out to find other ways to participate in the creation of knowledge and truth, some of which will be wrong or even deadly. In the future, we will need to address the ways in which people are marginalized in terms of the production and dissemination of knowledge, and the ways in which people react to that marginalization. The urgent need to recognize expertise and to evaluate information correctly in order to make the best decisions will be increasingly important, and we must address anything that thwarts that effort, like the aforementioned inequality and marginalization. The survival of a group depends on the group being equitable and fair to its members. When practices are unfair, people react in ways that can create problems for everybody. So if you want to survive, do what you can to be a productive member of a group that minimizes these negative reactions.

Practical skills, like making and repairing things, will be an important part of restoring systems that have collapsed. As an exercise, let us consider a world suddenly without electricity as an example to help us think of some of the skills that could be useful. These skills could be something as simple as starting a fire for heat, to cook, or for light. Knowing how to construct a stove to use fuel efficiently and knowing how to do this indoors safely would be useful knowledge. Understanding how electricity works and how to generate it from the sun, wind, or water could all be important skills, as would the ability to create batteries to store power. Knowing how to create an electric generator, a machine that converts kinetic energy into electricity, would be an incredibly valuable skill. Of course, we can live without electricity, but so many things are more efficient and easier when you have electricity as an option. We will not be starting from scratch, either. We are going to be in the same place, with many of the same things that we had before, but possibly without the ability to use them. Knowledge of refrigeration and air conditioning might be good skills to have, and it would improve our ability to store food or even our ability to live in certain parts of the world. Architects and contractors might be necessary to create buildings that work in the new reality.

The types of skills that we expect of our teachers will still be useful—mathematics, chemistry, and history will be necessary and desirable. Humanities and the arts will, too, although we do not always realize how much. When I worked and lived in the Pech community in Honduras, people there had many needs, including a water system, basic healthcare, and even transportation. More than any other single thing, though, the Pech wanted a video camera so that they could record dances, songs, and ceremonies—traditions they were afraid of losing. In their situation, faced with so many immediate and critical needs, they were concerned about their arts, their religious and cultural practices. Those sorts of things became more valuable, not less, in a long-term crisis. Skills in the arts will continue to be important. Storytellers, writers, painters, and musicians will continue to be important, as they always have been. When I think about the current pandemic, for instance, some of the things that have kept people sane are movies, video games, and music. These things do not address our immediate needs, like food, water, and shelter, but they are important nonetheless.

Ultimately, we will be part of a community. Our usefulness, and the influence that we can have with our particular skills, depend on our ability to function as part of a collective. Those with social and political skills might occupy a special place in the group, but basic traits like kindness, fairness, and empathy will be the foundation on which we build everything else. If there is sufficient disruption of our current situation, we might even discard the inequity of the present and start anew. In that way, perhaps, the next apocalypse can contain the seeds of an awakening, or a course correction.

SURVIVAL EQUIPMENT

Many of us who teach wilderness and preparedness skills would rather concentrate on what you know rather than what you have. You always have your knowledge, but you might not have your gear. That said, there are tools that can make it much easier to do what you need to do. Depending on the nature of the next apocalyptic scenario, we might not have access to all of the material things that we use in our normal daily lives. We might have little more than we can carry. Luckily (if that is the appropriate sentiment), there is an unprecedented interest in the kinds of tools and equipment that will maximize our chances of survival in a catastrophic emergency. "Survival kits," "bug-out bags," "every-day carry"—a litany of terms suggest our obsession with the gear that we might accumulate to stay alive and ready to handle an emergency. Our interest in survival gear parallels our obsession with apocalyptic narratives. We create postapocalyptic art, cosplay, and props. We make, find, modify, and sell the kinds of equipment our imagination suggests. We know what works in real-world situations, from our experience working or playing outdoors, or from first responders or the military. Second, acquiring and preparing gear, testing it, and packing it into our emergency kits allows us to do something about the potential apocalypses that we fear or about which we fantasize. Handling equipment makes us feel like we are active, not passive; we feel like we are taking control of our own future. In addition, preparing survival equipment is easy and immediate, compared to the hard and slow work of addressing the actual causes of our problems. If we worry about climate change, for instance, actions that result in meaningful and noticeable improvement are difficult and results are not instantaneous, offering no immediate gratification. Preparing an emergency kit, on the other hand, allows for instant gratification and serves to mitigate our feelings of helplessness. Third, we live in a world driven by consumption, and that world offers us new and better things all the time. We accumulate stuff. New and different products promise to be that much better, and sometimes they are. Prepping, disaster preparedness, and our interest in wilderness survival have spawned whole industries driven by this consumption. The final reason we spend such energy on our gear may be the most obvious: we might really need this stuff. Our survival gear may be necessary for survival, or at least for survival with a quantum of comfort. For many of us, surviving in certain conditions without some tools would be nearly impossible. For all of us, proper equipment can make it much easier and more tolerable.

What we need depends on the situation and the duration of the disruption. In my survival courses, I talk with students about what equipment we should keep in our emergency bag, or maybe in the trunk of our car. As my courses focus on surviving unexpected, short-term stays in the wilderness, the equipment reflects the needs we might have for a week or two. In that type of situation, lost in the woods, you need some way to stay warm and dry, you need water, and you need to be able to make yourself easy to find. In the short term, you can get by with little or no food. In an apocalyptic event, however, the duration of the disruption would be much greater, requiring different or additional gear.

The most basic toolkit should suffice for a situation in which you have to spend several days or a week outside. First, I would want a way to make fire, to keep warm or boil water. A close second would be some sort of shelter or the ability to make it. I would need the ability to find and carry water, and a way to purify it. A way to signal potential rescuers, like a whistle or a signal mirror, would be important. Although not always part of a survival kit or bug-out bag, I would want the right kind of clothing for the conditions, especially if it is cold. If you satisfy that wish list, you will have bought yourself enough time to find food, to signal rescuers, or to find your way home. Gear to assist in the fulfillment of those fundamental objectives might be important to us now, even before any kind of catastrophic event, if we lose some of the services that we take for granted, like electric power. This equipment would be doubly important in the event of an apocalyptic scenario, but only for a short while. Those are short-term survival needs, not a solution to the longer-term problem.

When I think of survival gear, or even camping gear, I always think of the time that our group ran into two men who were traveling through the rainforest in Honduras. We were documenting archaeological sites; they were hunting and fishing. I was traveling with four Pech from the village in which I lived, and we had been out in the forest for about two weeks. We saw the two men standing on a sandbar as we rounded a curve, headed downriver in a dugout canoe. As we always did whenever we encountered anybody, we stopped and observed from a distance. They turned and looked at us. Anywhere else, it would be weird, or funny perhaps—two groups of people staring at each other, 200 meters away—but here we wanted to know whom we were meeting, and if they could pose a threat that might make us consider a detour. The taller man had a rifle, probably a .22 from the looks of it. They both had on shorts and rubber boots, and each carried a machete. The tall one carried only that rifle, and the shorter man had a small backpack made from a flour sack, with ropes for straps. The four Pech I was with carried gear in the same kind of homemade pack. The short man also had a big bunch of green bananas slung over his shoulder. One of the Pech recognized the strangers, and could vouch for them, so we continued downriver and pulled the canoe up on the sandbar. We talked for a minute and decided we would all camp together that night.

The men had been out hunting for a few days. Everybody you would ever meet way out in the rainforest, except for us or some other scientific group, was hunting or fishing. The rifle was indeed a .22—an ancient bolt-action rifle. They had shot a deer with it and then chased it for a few hours but lost the trail. Many assault rifles from the Contra War in Nicaragua had found their way into the hands of folks in eastern Honduras, but only the wealthier cattle ranchers kept them, as ammunition was expensive. Most people did not have a rifle, or any firearm, and those who did had .22s. Shooting a deer in the head and running after it was standard practice, although it was ineffective and cruel to the injured animals who escaped.

We camped with the two men, and I saw what they were carrying when they laid it all out in the sun to dry. They had the clothes on their backs, rubber boots (no socks), a machete each, the .22 rifle, a handful of rounds, a sheet of tablecloth plastic that they used as a shelter, and a coil of rope (about twenty-five feet long). They carried two of the small cotton blankets sold in every little store throughout the countryside, a gallon jug for water, a cooking pot, two spoons, some matches, a half-pound plastic bag of salt, some fishing line wrapped around a wooden stick with a couple of hooks and lead sinkers, two spare D-cell batteries, an old metal flashlight wrapped in a plastic bag to keep the water out, and one flour-sack backpack. That was all they had, besides the bananas. We had some rice and corn meal, so we made rice and tortillas, along with a few little fish a couple of the Pech caught while the rice was cooking. It was good and we had plenty. Then we made coffee, and we had sugar for it. The newcomers were enthusiastic about the coffee, but they had everything they needed without us.

It would be so simple if only we could create a list of items to pack in a backpack that would allow us to survive and thrive during the next apocalypse. It is not that simple, clearly, but a good rule of thumb is that you must survive the first thirty days if you do hope to survive the next thirty years. Having the tools on hand to assist you in surviving during the first days after a catastrophic event is important, even if your needs change after that initial period. I created a minimal list of things everybody should have or be able to make. You do not need much, especially when it is warm. I will explain what the gear is for and why, and then I will give the specific example of what I use at the moment.

Fire

As I've mentioned earlier, before anything else, you need to be able to make fire. Your body temperature might depend on it, and you may need fire to boil and purify water. You could even use it to signal rescuers. Depending on your knowledge and experience, you might have a variety of means at your disposal. For the short term, you probably cannot beat a disposable lighter. Other things, like ferro rods (firestarter rods) to create sparks or lenses to focus the sun, will last a long time but require some experience to use in challenging situations. Unless you have a lighter, you will need good tinder that will start with a spark. Natural tinder can be found all over the place, such as the fuzz from a mature cattail, or dry leaves of certain types. Often, though, it is difficult to light natural tinder with a spark, and it can be frustrating. As I described earlier, cotton balls light easily with just the spark created with a ferro rod. Lint works in the same way, and other cotton things, like tampons or makeup-remover pads, can be pulled apart into a less dense, fluffier state and will act similar to cotton balls. Petroleum jelly can be smeared on the cotton and will enable it to burn for minutes, not seconds (although leave a bare spot to catch the spark). I create a fire kit with everything together in one place. Mine consists of a small waterproof container with cotton balls, a ferro rod and striker to make sparks, some Vaseline, and a disposable lighter.

Knife

I think a knife is one of the most important things you can have in short-term survival situations. Cutting instruments were among the first tools that humans ever created. You can make most of the rest of the kit with knowledge, skill, and a cutting tool. The ability to cut things or to shape things is fundamental to many of the tasks that we need to do, so a good knife is important. Something like a pocketknife would be better than no knife, but not as good as a larger lock-blade folding knife, or better yet, a medium-size (four-to-five-inch) fixed-blade hunting or survival knife. A larger, more robust knife will be far more useful than a small pocketknife, although heavier, harder to conceal, and sometimes not as good for fine work. In some cases, another tool might do the work of a knife. Folks in rural Honduras use machetes for everything, from preparing food to clearing a path, but machetes require significant experience and practice to use them competently and not hurt yourself. Deciding which knife is the best kind depends on the intended use, but some general rules apply anywhere. The first rule is that simpler is better. Fewer parts to break usually means that things will break less often. A fixed-blade knife has less to break and is probably a better choice than a folding knife. How big of a knife you want probably depends on the task, but we can think about the kinds of things you do around the house and in the kitchen already. You want a knife the size of those you use in the kitchen to chop vegetables or cut up a chicken. In fact, those very kitchen knives might be fine. There are certain things that make some knives tougher than others, such as the way they are constructed. One important feature is the way the handle is attached to the knife and whether the metal to which the handle attaches (called the tang) extends all the way to the end of the handle. A full-tang knife, in which the metal of the blade extends as a solid piece all the way through the handle, is desirable because the blade will be unlikely to snap off the handle. A full-tang, fixed-blade knife with a blade around four to five inches long would be a useful and versatile knife.

Knives are made of many different metals, and everybody has an opinion about what is best. Some keep an edge but are harder to sharpen, and others are just the opposite. Stainless-steel knives will not rust. Then there's carbon steel, which can rust but often takes and keeps an edge better. I have some of each, and some are better in certain situations, but most of the time, anything works. I prefer stainless steel for marine or wet environments; otherwise, it makes little difference to me. If you're doing things like chopping wood, it would make sense to use something like an ax, a hatchet, or a machete, but you could probably also do it in a pinch with a decently sturdy knife. In the kind of postapocalyptic scenario that I am imagining, where people are in and around communities as opposed to being lost in the woods, you might have access to a variety of tools, so you don't have to try to do everything with just one knife.

That said, when I lived in Honduras, nobody carried a knife except to clean fish. They did everything else with the machete. In fact, most people carried only a machete when we went on our extended trips into the forest, and they did everything with it. Whenever we were out in the field, the machete cleared the trail, then the campsite, and later cut up the chicken and vegetables for dinner. Back at home in the village, people had a variety of tools, and would use those instead. Even at home in the village, however, a machete was still useful far beyond what many of us envision. Worldwide, people use a variety of different tools for cutting, and you might prefer one over the other depending on what you are accustomed to using. Many solutions work.

I have many knives that I have tried and that I let students use in survival courses. I go back and forth between a few that I like best, because of how they work and how they feel in my hand. Currently, my favorite is a Ka-Bar Mark 1 knife, a fixed-blade knife with a five-inch blade that is a reproduction of a general-purpose knife used by the U.S. Navy in WWII. I also like the ESEE-4 knife a lot, which is a slightly smaller fixed-blade knife. A very inexpensive knife might be weak or made of inferior steel, but there are decent knives out there for very low prices, particularly from big companies like Morakniv or Gerber. If you are careful with your knife, any hunting or outdoor knife should be sufficient. The knife I use most is a little pocketknife, a yellow Case Sod Buster Jr. that I usually have with me and use to open packages and for other everyday tasks. That would not be ideal in the wilderness but would work if that were my only choice.

Water

Water will be important, and having some on hand will be useful. You will eventually need more than you have, though, so you need to be able to collect it, purify it, and carry it. Waterborne illnesses like diarrhea are annoying, but they can also be dangerous because they sap your strength and dehydrate you. Water purification is essential. Some kind of water filter might be the easiest way to do this, but boiling water is a sure thing. In order to boil water, you will need some kind of container that will not melt, and usually this means a metal water bottle or pot. You can boil water in a thin plastic bag (the water wicks away the heat from the plastic before it reaches melting temperature), but almost any other container would be a better option. A single-walled metal water bottle is a good option—you can carry and boil water in it. The double-walled bottles, insulated to keep things hot or cold, are not good for boiling water, since they are designed to limit heat exchange. If you have a metal pot, metal cup, or military canteen cup, you can use that for boiling water, allowing your water container to be of any material. You boil water in your cup and then transfer it to the bottle.

I carry a military-style canteen and canteen cup usually, but sometimes just a disposable plastic water bottle that I keep refilling and a metal pot or cup in which to cook. I usually take stainless-steel cups because aluminum ones get crushed in transit too often. Titanium is an option, lighter and tougher than anything else, but it is expensive and I have never thought it was worth it.

Shelter

Tools to create a shelter would be next on my list. Found shelter would be the easiest, like a car, airplane, or building. Outside of that, a tent might be the easiest and most obvious thing, but a tarp works as well. A lot depends on the conditions. In the tropics or in the summer anywhere, you will need less protection than in cold weather. In an area with high winds, your shelter might look very different from what I might make here in Kentucky, where winds are typically low. If you do not have a tent or tarp, many items can function as makeshift shelters, including piles of leaves, cardboard boxes, and sheets of plastic. Since the obvious solutions to most problems are easy to figure out (like carrying a tent for shelter), I tend to focus in my survival course on the unexpected ways in which we can fashion expedient tools in the absence of an obvious choice.

For shelter in a survival kit with limited space, I take a military-style poncho or a small tarp, maybe 2 by 2.5 meters (six by eight feet) or so and some string or cord to tie it up. I carry a few metal tent stakes, but you could make those from sticks if needed. If I were camping, I would have a regular tent, which is much more comfortable than a tarp. For years, I carried a hammock or tent with a rain fly and mosquito netting. I used Hennessy Hammocks in the rainforest, and those were great because you could camp on a slope (hanging between trees, the slope of the ground did not matter), and hammocks are much cooler than tents. I like tents, however, because you have room to spread out your equipment and to store it inside. I do not have a favorite tent, though. Any tent works, although some are more durable or lighter than others, and some breathe better (avoiding condensation inside) or shed rain more readily. Usually these features are reflected in the cost. I would choose durability over weight savings in most circumstances.

The choice of blanket or sleeping bag depends on many factors, and the tradeoffs are familiar to most of us. A down blanket or sleeping bag is warmer per weight and volume when dry, but not effective when wet. Wool is heavy but tough, and better than most other things when wet. I usually end up with a blanket or bag with synthetic insulation. Military poncho liners are popular for use in mild climates and are relatively inexpensive. Most of my experience has been in the lowland tropics, where a light blanket sufficed, but in the limited cold-weather camping that I have done, I used a down sleeping bag. Thin mylar emergency blankets can add warmth when layered with a blanket or sleeping bag, or course, but they don't breathe, and condensation can be an issue. If you think of this in terms of layers, as you would with clothing, you can create a flexible and sufficient system from a few different components: a thin blanket, something thicker, and a mylar emergency blanket that can be combined in various ways as conditions demand.

A lot of heat is lost to the ground, so part of your insulation should involve a sleeping pad of some sort. I used a self-inflating Therm-a-Rest mattress for years, but I have also used thin foam pads. There are many light, inflatable sleeping pads now that are actually comfortable, but I have no experience with those for long-term use and I do not know how durable they are. In the tropics, I slept in a hammock with no pad or in a tent with leaves piled under the floor, because losing heat was not a problem.

Medication

One of the last essential pieces you must consider is any medication you need. Getting more than a few months' supply of medication in advance can be challenging, depending on the circumstances, so this is a short-term solution. If supplies are not available, the only option left might be to look for a natural alternative, some change in diet or behavior that helps with your condition, or something that can be manufactured in the new reality. The need to get systems reset becomes apparent when we think about a situation in which we cannot access lifesaving drugs. Homemade alternatives might provide good options to treat conditions in some cases, but not all.

Light

At this point in the list, we have moved beyond the life-or-death items. A light source at night—either a flashlight or headlamp—makes life a lot easier, but you can survive without it. We all have lights on our phones now, but a little flashlight that fits on a keychain is cheap, will always be with you, and the battery can last several nights, even weeks, if used sparingly. In a backpack, I usually carry some sort of headlamp rather than a handheld flashlight so that I can have my hands free to carry things, cook, or wash dishes. Water-resistant headlamps are something I looked for in the rainforest, although you could always put one in a plastic sandwich bag during the rain. Almost all lights sold today are LEDs, which I prefer because of their greatly extended battery life. I have used everything from relatively expensive headlamps from a camping store to cheap headlamps from the hardware store. They all worked and none failed. I suspect the price might reflect toughness and durability. In my car, I have a cheap flashlight that can be hand-cranked for power. While usually providing weaker light than battery powered ones, crank flashlights turn kinetic energy that you supply into power and would be good for a long-term situation in which batteries could not be replenished.

Backpacks

Last, or maybe first, you will need a bag to carry all of your gear. A backpack is easier to deal with than a duffel bag or shoulder bag if you are walking. I like some sort of pack with no or few zippers, as those can fail and make it hard to close the bag. Cheap and tough examples include military rucksacks, although they are heavy and some people may want to avoid a military appearance. Military ALICE (all-purpose lightweight individual carrying equipment) packs are ubiquitous and cheap, and the smaller of the two sizes can be used without a frame. In Honduras, the Pech would make backpacks from rice bags and a length of rope. I carried one of these for a few days, and for relatively light loads (less than thirty pounds) even that works fine. The most common failure point I saw on any backpack, even the cheapest ones, were zippers and plastic fasteners. In no case did I witness a failure that caused anything more than a minor inconvenience. I saw a hole worn through the fabric of a high-end, lightweight backpack belonging to a student during a daylong truck ride in Honduras where the bag was tied to the roof of the truck, rubbing against something, and I saw a similar pack ripped by a thorny vine in the rainforest. For me, the extra weight of heavier fabric is worth the tradeoff when it comes to durability.

In some ways, this whole list looks a lot like what you might take camping, and it is. There are a few differences, though, between camping gear and survival gear. When we are camping, if we are hiking in to our campsite, we may be more concerned about how much weight we are carrying rather than the durability of our gear, and in a survival situation that might be the reverse. You might not choose the ultralight backpacking gear made with thin nylon for a longer-term and unpredictable survival situation. While sufficiently tough for hiking and camping, some high-end camping equipment is not nearly as durable as a cheap, old-fashioned thick nylon or canvas rucksack. That heavier gear would be overkill for a through-hiker on the Appalachian Trail. The two situations are different. When I was working in the rainforest in Honduras and we would be out for two or three weeks at a time, I ended up using a lot of military surplus equipment, even though it was heavy and old-fashioned. I used it because it would not break in the middle of a trip. For instance, I much preferred carrying an ALICE pack to a fancier, lighter commercial backpack. It had no zippers to break, it never ripped when it caught on a thorn or on a piece of metal, and it was never damaged when strapped to the back of a mule. Its simplicity and toughness became the most important qualities to me. Similar principles carry over to practically everything else. You want simple, tough shoes and clothes, things with the least number of zippers and buttons or other things to break. In that sense, picking equipment for some sort of survival situation or a long-term use, you would make different choices than you would make in a recreational setting. In general, though, both situations demand similar kinds of equipment. If you have some camping gear, you probably have the basics for a survival kit.

One other difference between a survival situation and camping would be the importance of getting information and communicating. When camping, you might want to escape the constant connectedness that we enjoy (endure) in our daily lives. In an emergency, however, getting information and communicating with people may be much more important and urgent than during a camping trip. A small radio to keep informed of what is going on could be critical in an emergency. In addition to AM and FM bands, you can get radios with shortwave bands and weather radio bands. A cell phone or walkie-talkies might be necessary to communicate with your family or other members of your community, and those might be part of your preparation.

Self-defense

I am of two minds about what you should know or do in order to protect yourself from threats. On the one hand, I think it only makes sense to have the skills to defend yourself and to have weapons or tools to protect you, your family, your community, and the stuff that you need to survive. On the other hand, as a strategy for a long-term situation, it does not seem tenable to stay in a place where self-defense is regularly necessary. Imagine this situation: You have weapons and training, and you can use them effectively to protect what you have. Imagine that you have a month's worth of food, and people are trying to take it from you. Now, in this situation, those people will also be armed and trained. Whether we imagine knives, machetes, or firearms, we are talking about a potentially deadly encounter. Now, let us assume that you are far superior in this type of situation than the people you will be facing. You have trained, you have good equipment, and you are fighting from a fortified location. You are going to win that fight 95 percent of the time. You will win 19 out of 20 times. On average, however, you will have lost after about ten conflicts. If this happens once a month, you'll have a 50 percent chance of having lost in less than a year, and in less than two years you are almost certain to have lost one of these deadly encounters. If we change the frequency, the loss could come much sooner. If it were happening every week, then you would not last six months. We can change all of these variables, but you are going to lose, eventually. In this calculation, I am not even considering the toll that the stress of repeated, life-threatening encounters would take on a person, a family, or a community. You cannot stay alive like that. Even if you could, you cannot live like that.

Any of us who have lived in places with rampant gun violence recognize this scenario as consistent with what we have experienced or have heard from those who lived it. In doing research for the book *Jungleland*, author Christopher Stewart and I took an extensive trip through Honduras, traveling along the coast and in the rainforest for about a month.[7] We talked to at least a dozen families who had relocated to the edge of the rainforest, and we asked them why they had done this. In every single case, they had moved to escape violence somewhere else. Every single person we talked to was fleeing violence. The situations they fled looked something like this: a member of one family would have a problem with a member of another family, violence would escalate, and somebody would be killed. There would be retribution for that death, and then retaliation from the other side. The cycle will continue. The people who fled the violence in Honduras were not the losers of these conflicts. These were not people unfamiliar with conflict, or unarmed, or pacifists, or liberals, or some other category that we tend to think of as unable to deal with this kind of situation. They were exactly like the people with whom they were in conflict. They were armed, they were proficient with weapons, and they were tough and resilient. At the end of the day, however, the cost was just too high. When your opponent has lost all of their children, and you have lost all but one, you would not consider yourself a winner. Nobody is going to fight a war of attrition with her own family, which is exactly what was happening. In the same way that the cost was too high for those people to stay where they were and continue this losing battle, that kind of strategy is not going to be sustainable in the aftermath of a catastrophe.

I am not suggesting that you should not protect yourself or your stuff. I am saying that if you are in a situation where your continued well-being depends on you having frequent high-risk violent encounters, you are bound to lose eventually, and that is not a viable strategy. You are going to have to figure out how to create a system where there is enough to go around, and where you are not at risk for that kind of violence. As always, I keep coming back to community as the solution. You cannot go it alone and you cannot hoard things, because eventually people in desperate need will come to take what you have, and you cannot stop them. You cannot live like that, even if you win at the first ten or twenty times. Any situation that requires the kind of self-defense and violence that we find in our popular narratives has to be very short-term, and a solution to the violence must be found quickly.

INTELLECTUAL SKILLS

Recognizing Competence

Just before the pandemic hit in early 2020, I sat with archaeologist Scott Hutson in a bar in Lexington. We had talked about "collapses" in the Yucatán Peninsula of Mexico, and how the focus on collapse threatened to oversimplify and obscure a complex reality that required subtlety and attention to detail to understand. Turning the conversation to potential future collapses, and the types of skills that would be useful, Scott quickly identified what he considered the most important things we must do: recognize competence in leaders and experts and know how to assess and analyze the information we have to make sure that we, as a group, are headed in the right direction.

During the pandemic of 2020, the cost of failing to recognize competence and listen to good advice was apparent and visceral. People were dying, and we saw the ignorance and stubbornness in the refusal to practice social distancing or to wear masks in public—both shown, without any doubt, to have a great impact on the number of cases of Covid-19 in a given area. The distrust of expertise, the politicization of knowledge, and the inability or unwillingness to acknowledge reasonable advice from experts led to the United States becoming a low-performing, near-pariah state, perceived by others as dangerous, in rapid decline, and to be avoided. The European Union decided to prohibit most U.S. citizens from traveling to Europe in July of 2020, and the subsequent prohibition on U.S. citizens traveling anywhere except nine countries with no restrictions (out of 195 countries worldwide), and another thirty-five with restrictions. Eventually, with the vaccine rollout and a change in leadership, the situation improved. In any crisis, knowing who has good, reliable information will be critical.

Evaluating Information

A correlate to recognizing confidence, or perhaps another aspect of that same phenomenon, is the ability to evaluate information and think critically. Although this sounds like something from a pamphlet for a liberal arts college, critical thinking was the first skill mentioned by literally every single person I spoke with about apocalypse scenarios, from college professors to artists to wilderness skills teachers. In order to make good decisions, not only do you need to know where to go for information (recognizing competence), you need to be able to assess information. Another phenomenon we observed during the Covid-19 pandemic involved how people reacted to information and data presented to them. These responses reflect the types of logical fallacies and other common mistakes that hinder a critical analysis of information. In reviewing survival situations, I find that the ability to distinguish correlation from causation to be particularly important, along with the ability to make decisions when the data is incomplete or contradictory. Colin Powell (former secretary of state and general in the U.S. Army) writes about leadership in his widely shared PowerPoint deck entitled *Leadership Primer* from 2006, in which he delineates eighteen rules of leadership.[8] Number 15 is the "40–70 rule," in which we are advised to collect between 40 and 70 percent of the information we need to make a decision. Less than 40 percent, Powell has decided, would not be sufficient information. Delaying a decision in order to obtain more than 70 percent of the information you might possibly collect would make many decisions untimely and too late to be effective. I might paraphrase this rule as "collect enough data to be reasonably certain, and if time is a factor, go ahead and make a decision." More time pressure might force you to make a decision closer to the lower end of data collection, while more time might allow the luxury of collecting more.

So many factors go into collecting and assessing information that writing about it in a general way is difficult. Having basic knowledge of your surroundings, of how things work, and of what you need to survive would be a good start. Understanding common logical fallacies and basic statistics would be a good next step in preparation, as would an understanding of how and when to use anecdotal data.

POLITICAL SKILLS AND SITUATIONS

Flexibility and Adaptability

As seen in Part 1 of this book, archaeologists almost reflexively recoil from the term "collapse" these days. Part of that is a reaction to publications like Jared Diamond's book *Collapse*, in which the focus on "collapses" was seen as ignoring resilience, and in which his identification of the causes and mechanism of "collapse" seemed faulty or simplistic. But another part of the negative reaction to the term 'collapse' comes from a recognition of the ways in which collapse narratives end up blaming the victim. By speaking of collapse, we imply that a society somehow failed—at the very least, failed to continue. Most archaeologists now look at those situations and see flexibility and adaptability. While some things end, other things persist, and this continuity can be more impressive and important than the disappearance of certain elements from the archaeological record. Examining the past from a different angle, focusing on the resilience of ancient people rather than the collapse of a certain segment of the ruling group gives us a blueprint for understanding what we will need in the future: flexibility and adaptability.

In a recent study looking at the relief efforts after Hurricane Maria in Puerto Rico, information technology professor Dr. Fatima Espinoza found that community leaders and activists did not try to re-create the structures that had been in place before the hurricane. Rather, mindful of the inequalities and failures of that system, and being experts in their own communities, they used new technology, as well as a mix of people and organizations to meet the goals they set out to achieve.[9]

I spoke about this topic with Dahlia Schweitzer, the author of *Going Viral*, in which she analyzes outbreak narratives. Schweitzer thinks that our social infrastructure, our social safety net, would be critical to survive upcoming disasters. "I think the most pressing things that we should have in place, yesterday, is universal healthcare. I think this [Covid-19] outbreak has just been one more item on the list for why we're kind of screwed with healthcare. Then also, we have the people who are getting laid off, so they lose their healthcare. Or they can't afford to take sick time off work so they're going into work anyway and they're spreading the virus. On so many levels, I think universal healthcare. And then, you can even extrapolate from that to social services in general."

Any recovery effort from a future apocalyptic event would benefit from the lessons learned during recent catastrophes. The ways in which insufficient systems of healthcare, housing, education, and food security exacerbated the disaster suggest the importance of working at this larger, structural level to create a community in which everyone's basic needs are met. Leaving people out eventually costs everybody.

CONCLUSION

"So what is your escape plan?"

We were eating dinner when my friend asked me this. She had taken an urban survival course with me and was worried that climate change and the current political situation could precipitate a crisis.

"Where are you going to go when it all goes wrong?" Her question sounded offhand and spontaneous, but I knew this was what we were meeting to discuss. Her question was serious, and she was worried about the near future. She wanted some advice, and some hope. I know where she hoped the conversation would lead. I must have a plan, she reasoned, that would allow me to utilize the wilderness survival skills I teach to flee the city and make a life in the hills. Or maybe, because I am an anthropologist, I knew which part of the world would be spared the worst of it.

I told her my plan was not to escape: it was to stay and help. If it all falls apart, there will be great need, and I will try to find a place I can contribute. There won't be any escape. There are too many people, and the only future worth living requires us all to stick around and solve the problems together. Media depictions of the apocalypse are very different, and these fantasies can hamper our ability to plan for and react to a more realistic scenario.

In the past, consistently, we see people reorganizing, regrouping, and creating structures and systems that allow the new community to persevere. That was not the quick solution my friend was hoping to hear, perhaps, but she smiled, and I could see in her eyes that she already knew this was the answer. She herself was a community activist and organizer, and nothing about her life was selfish. Part of her discomfort, I imagine, came from the way popular narratives represent survival. Everything she was currently doing was for her community. Surviving by fleeing, by abandoning her people, was not really an option. When she thought about an apocalyptic catastrophe, she defaulted to the reactions we see all the time in apocalypse stories, where you grab your bug-out bag and hide out from all the unprepared folks. She was never going to do that, and when I told her my plan, it all clicked. When trouble hits, you head for the people who might need your help, and who will help you in return. The skills we had covered in the survival class could come in handy, of course. They were temporary, though, like those miniature spare tires: they'll get you home, but you don't want to drive on them for long.

Our apocalyptic fantasies and fears make it clear that some of us want to escape what we have now, while others fear losing it. Sometimes we want to do both at once. Some of us welcome the changes we imagine, and others dread those same changes. It might be that most of us fall somewhere in between. Our ideas of an apocalypse are not very plausible, and our means of preparation seem geared toward the short term rather than the long term. From all of this, we can draw the following conclusions.

First, our vision of the future shapes the future. We already react to things every day in ways that are informed by our expectations. We create our future; it does not just happen to us. The event that sets off the chain of reactions that leads to catastrophe might have nothing to do with our vision of the future, such as a supervolcano erupting or a comet hitting the Earth. Even in those cases, however, our response is not necessarily predetermined. What preparations we make, and how resources are distributed shape the aftermath. There are no natural disasters, just natural hazards. Our response matters.

Our fears and desires form our vision of the next apocalypse, and that vision is unlike any of the apocalyptic events we see in the archaeological or historical record. Rather, our apocalyptic fantasies reflect the contemporary world. Some of our fantasies reflect deeply problematic attitudes, including racism, sexism, and xenophobia. They do not prepare us for the future.

I see some lessons from the current pandemic and from recent natural disasters that should affect how we envision a future apocalypse playing out. First, few of us anticipated how political the response to truly tragic catastrophes could be. The Trump administration's incompetent and negligent handling of the Covid-19 pandemic would have seemed nearly unbelievable a few years ago, as would the degree to which our responses to the pandemic became political symbols. Such heinous self-interest was something I associated with the barbaric and brutal leaders of the distant past, or contemporary leaders of an authoritarian regime. The self-defeating response of many members of the public, refusing to take basic safety measures, or to understand the true nature of the Covid-19 virus, was not something most of us predicted. As Dahlia Schweitzer told me in our interview, no Hollywood screenplay would have seemed realistic or plausible with that kind of plot.

We saw the same kind of politicized response to hurricane Maria in Puerto Rico, again by the Trump administration. Similar examples are found worldwide throughout modern history. It takes various forms, from the misuse of resources meant to help people through the disaster to the refusal to acknowledge a crisis for fear of geopolitical ramifications. Envisioning the next apocalypse requires that we leave a space for the type of government and leaders we have in place. In some systems, the leader makes a great difference. In other cases, the individuals involved might not have as great of an influence as the type of political system. In the United States, we see that the public's response varies greatly by political party and by the competence, empathy, and ethics of our leaders.

An important revelation from work like that of Fatima Espinoza on the response to Hurricane Maria is that official, institutional responses are very likely to be lacking for a number of reasons, from a general lack of competence as we saw in the Trump administration to preparations that do not adequately account for the nature of the catastrophe. It's better to plan on an inadequate governmental response, even if those systems remain intact.

I look at the past to envision the future. We have always looked at history to understand the present and to anticipate the future. Those who do not study the past are doomed to repeat it, the saying goes. Archaeologists understand that our research has implications for the present, and we talk about the relevance of our research to modern, contemporary society. Looking at the past to understand the future is nothing new. Looking at historic or archaeological examples of catastrophes is particularly important because we are currently creating the narratives of the next apocalypse. These stories influence our expectations for the future, and many of them are at odds with what we see in the past.

While being cognizant of the many significant differences between any two places and times, we can still learn lessons if we look at the commonalities. In the examples I've explored here, we see that collapses are the culmination of long-term processes. An apocalypse is likely to be multicausal, with various things that set it off. A single event or series of events may set it in motion, but the entire network of systems and structures in which we live becomes involved quickly. Realizing this is important because preventing a collapse or recovering from one will require multifaceted responses, at a number of scales, and in many ways beyond those that deal with the proximate cause.

One last reason to look into the past is so that we understand how it influences our narratives of the future and our view of the present. By making explicit the content of those narratives, we can understand what they signify. For instance, we see that many of our apocalyptic narratives contain strong patriarchal and sexist content. By examining our narratives, we can identify attitudes and elements of the status quo that we do not want to perpetuate in the future. It helps us see how our longing for a mythical past that never existed can shape our vision of a future that will be. We need to understand that our visions of the future are not based on some reasonable examination of the evidence, but rather on wishful thinking.

The past is not like the present or the future. A number of important differences exist, including a few key differences that will determine all the others. The number of people who will survive nearly any catastrophic situation we imagine today will be much greater than during any of our past examples. In our fantasies, we end up in small groups, with equally small needs. In reality, we will have entire communities of people to feed, house, and provide the basic necessities for. In no scenario do we approach the small, low-density populations we saw before the near-universal adoption of agriculture by three thousand years ago. This means we will need a viable agricultural system, or systems.

I said at the outset that this is not a doomsday book and that I am not interested in fearmongering or advocating for some simpler, traditional way of life. However, I do not want to downplay what I think will be very difficult changes in our future. I think the next apocalypse will be brutal, tragic, and costly. We will wish it had never happened, and the loss will be enormous in terms of lives, livelihoods, material possessions, and ways of life that we value. I think climate change will be the proximate cause of many near-term catastrophes, and we should not understate the severity of the impact of climate change that we have been ignoring for fifty years or more. It will be awful and truly catastrophic.

Our response is part of what creates a catastrophe. Rebecca Solnit documents the extraordinarily positive ways in which people react in the immediate aftermath of disaster.[1] I wonder if that would be a relatively short-lived response, however. I am not sure if our better nature will always prevail over a longer period. We should be concerned about illiberal, reactionary thinking in times of stress, as we saw in the 1930s in Europe and more recently in the United States and elsewhere. People do not become progressive, informed, fair, and equitable after they go through a crisis. Often, they look for somebody to blame, and the blame will fall on the least powerful, the marginalized, and those unable to defend themselves. This is one reason to examine our fantasies. In order to guard against an oppressive response to a crisis, we need to identify the seeds of those tendencies in the narratives we have already created. We will need to fight against that. If we know it is coming, if we can check ourselves when we start to fall into the easy traps of scapegoating or looking for simple solutions that come at the cost of others, we have a chance to re-create ourselves in a much better way.

Even though we will confront a crisis as a community, there are things we can do as individuals to prepare for or prevent the next apocalypse. Almost everybody I interviewed for this book suggested that one of the most important skills in the future will be the ability to make good decisions, including evaluating information, choosing good leaders, and knowing whom to listen to for information and advice. In many ways, this is the most important skill we can develop now. In terms of preventing the next apocalypse, we have to know what is really happening, even if they are inconvenient truths. We need to realistically assess our ability to do anything about potential crises, and we must understand what to change and what it will cost to minimize the likelihood of the next apocalypse, or to mitigate the negative impact of future events. We must educate ourselves to identify and listen to the right people, choose the right leaders, and be able to assess our situation realistically. Viewing a crisis through a partisan lens, a vengeful lens, or with fear and trepidation will only increase the magnitude of a future crisis.

In any apocalyptic future, we will all have to change in many uncomfortable ways. We should prepare ourselves for that eventuality. To minimize the changes to come, we may have to change in uncomfortable ways right now. The level of discomfort of voluntary changes made now might be much less than the discomfort that could result from changes that would be forced upon us later, during a catastrophe we might have avoided.

Preparation by acquiring the types of bushcraft skills that I teach could be useful, even if in a limited way and for a short time. These skills are easy to learn and they cannot hurt. Learning how to do as much as possible with minimal tools or technology may allow us to shape a future that avoids some of our contemporary problems related to energy use and excessive consumption. Learning as much about engineering and technology as possible could help us mitigate some of the losses. Learning these skills and acquiring this knowledge *can* have a real effect in that they can boost our confidence, reduce fear, and allow us to keep clearer heads when making decisions. Therefore, while I think we might be unlikely to use these skills in a remote wilderness situation, they might have a positive net effect. Given the relatively low investment of time to learn them, acquiring some basic bushcraft skills might be a cost-effective way to prepare for any eventuality.

When we think about preparation, we often think about gear or equipment that we can purchase. Preparation has become a commodity, and someone is always there to sell you the thing that will get you through the next disaster. As we explored, creating a basic emergency kit is a good idea for both small-scale emergencies and for something truly apocalyptic. Most of the items are probably already available in most households, like a knife or a tarp, and creating an emergency kit should be possible on nearly any budget. I have created survival kits for my family, and I have various tools easily accessible. I suspect, however, the usefulness of these items will be limited to the first part of what would prove to be an ongoing process.

When we think about rebuilding a society, we must ensure we have the collected wisdom from our past. We do not want to reinvent the wheel. We must ensure that we have knowledge accessible in an uncertain future. We will need a durable, non-electronic version of the information necessary to rebuild. The fact that so much information now is digital and online, and that accessing it requires certain equipment, infrastructure, and electricity should give us pause. To get around this problem, we might need to ensure we have hard copies of these things, including textbooks, scientific journals, and encyclopedias. Information backup might not be first on our list of things to do, but it's worth thinking about.

Another way to preserve knowledge is to preserve the *people* who have this knowledge. We do that now by teaching and mentoring students. It is clear, however, that the people engaging with some of this knowledge come from a very narrow segment of society. Only certain people, and certain types of people, are pursuing certain careers, such as agriculture. As corporate farming increases, fewer people are participating in agriculture. In places like the United States, the kind of person who participates is relatively homogenous. Less than 1.5 percent of the population in this country engage in agriculture, and most are rural white people. Fewer than 2 percent of farmers are African American, and around 3 percent are Latino.[2] Many minority farmers are located in particular geographic zones, such as African American farmers in the South. Some people try to portray a concern for diversity as some politically correct, problematic handout that favors certain groups over others. No one who seriously engages with diversity efforts or studies the effects of it believes anything of the sort. We know that diversity benefits the system being diversified, not just the people who might have an opportunity to participate. As an archaeologist, I am limited in my ability to understand the past and the present. Everything from gender to socioeconomic status shapes the way I experience and understand the world. This may not be immediately obvious. As a white man in the United States during the twentieth and twenty-first century, I could easily ignore race and gender as factors influencing the trajectory of my life. I could delude myself into believing that I pulled myself up by my own bootstraps, and that my journey represents some sort of meritocracy, where you finish where you deserve. It is not that way, of course, and we all know this. The limitations in our knowledge, vision, and understanding that result from a lack of heterogeneity or diversity may be hidden, but they are real and we will need to address and minimize these limitations in any postapocalyptic future. One thing we can do to prepare for the future is to ensure that a diverse set of people spread across the globe have the knowledge required to reconstruct and maintain the systems we depend on. We do not want all the farmers in the United States to be located in the Midwest, for example. Perhaps, had we kept this in mind all along, we would currently have a greater diversity of viewpoints, ideas, and approaches that would have enabled us to avoid the next apocalypse. Perhaps it is not too late.

As I discussed at length, the narratives that we create shape our vision of the future. Those narratives are informed by our fears and fantasies of the present. Therefore, our vision of the future directly relates to the conditions in the present, it affects how we plan for the future, and how the future plays out. It is a big feedback loop, and we need to understand that. If there is a singular takeaway, it may be this: our fantasies and preparation for the future are not preparing us for the next apocalypse. An examination of the past shows us this. What we need cannot be commoditized, or put in a backpack, and the skills we need are not ones you learn in a weeklong wilderness survival course. Communities will survive the next apocalypse, not just individuals. Leaders might emerge, but heroes are not going to solve our problems. Heroic acts will be community focused, and may consist of a thousand subtle, unremarkable actions that we undertake for the duration of a crisis. Heroism may not be very cinematic and flashy, but rather sustained, quotidian, and anonymous.

Apocalypses happen, all the time, on varying scales. Surviving them, like surviving most things, is a community effort, not an individual effort. You will need other people, and they will need you. The skills you bring, and the ones you learn along the way, will help the group. Ultimately, that community-mindedness and altruism will generate strong collectives. That is the only way we can persevere. That is what survival looks like.

ACKNOWLEDGMENTS

We will not survive an apocalypse on our own, and I could not have written about it without the help of a community. Above all, I want to thank my wife, Soreyda Benedit Begley, whose unique perspective shaped mine. She shared the burden of completing this project during the pandemic while navigating the challenges of a full house and new obligations for both of us. I am grateful for her, and for the constant companionship of our children, Bella, William, and Aaron. They endured a house full of apocalyptic films, survival handbooks, and constant talk of dramatic societal change while living through it.

Thank you to the archaeologists who taught me, beginning with Bill Sharp, Steve and Kim McBride, and Gwynn Henderson in Kentucky, and George Hasemann, Gloria Lara Pinto, Pastor Gomez, and Boyd Dixon in Honduras. I am grateful for my recent colleagues, underwater archaeologists Peter Campbell and Roberto Gallardo.

My old friends Walt McAtee and Cliff Westfall helped shape this book and endured hours of conversation on this topic. I appreciate their time and effort. Their comments on early drafts improved the subsequent versions immensely.

I want to thank all of the people I interviewed for this book, including Ricardo Agurcia, Catherine Besteman, Heath Cabot, Craig Caudill, Kara Cooney, Josephine Ferorelli, Alejandro Figueroa, Gwynn Henderson, Scott Hutson, Takeshi Inomata,

Patricia McAnany, Guy Middleton, Riccardo Montalbano, Adam Nemett, Chris Pool, Karen Ritzenhoff, Dahlia Schweitzer, Bianca Spriggs, and Mantha Zarmakoupi. These conversations shaped my thinking about the next apocalypse and facilitated the presentation of their complex ideas in a clear, colloquial manner. I especially want to thank Kara Cooney, Bianca Spriggs, and Dahlia Schweitzer, who not only spoke with me for this project but also continued the conversation for my radio show, *Future Tense*, about life during and after the pandemic.

Finally, I want to thank Leslie Meredith, my literary agent, who reached out and suggested this project, T. J. Kelleher, my editor at Basic Books, and copyeditor extraordinaire Rachelle Mandik, who helped guide me in engaging a broader audience than in my academic writing.

BIBLIOGRAPHY

Aimers, James, and David Hodell. "Societal Collapse: Drought and the Maya." *Nature* 479 (2011): 44–45. https://doi.org/10.1038/479044a.

Aimers, James J. "What Maya Collapse? Terminal Classic Variation in the Maya Lowlands." *Journal of Archaeological Research* 15, no. 4 (2007): 329–377.

Aldiss, Brian. *Billion Year Spree: The True History of Science Fiction*. New York: Doubleday, 1973.

Ali, Safia Samee. "Where Protesters Go, Armed Militias, Vigilantes Likely to Follow with Little to Stop Them." NBC News, September 1, 2020. www.nbcnews.com/news/us-news/where-protesters-go-armed-militias-vigilantes-likely-follow-little-stop-n1238769/.

Aradia, Sable. "Sexism in the Apocalypse." *Between the Shadows* (blog), Patheos.com, September 1, 2017. www.patheos.com/blogs/betweentheshadows/2017/09/sexism-in-the-apocalypse/.

Arnauld, Marie Charlotte, Chloé Andrieu, and Mélanie Forné. "'In the Days of My Life.' Elite Activity and Interactions in the Maya Lowlands from Classic to Early Postclassic Times (The Long Ninth Century, AD 760–920)." *Journal de la Société des Américanistes* 103 (2017): 41–96. www.jstor.org/stable/26606790.

Baker, Brenda J., George J. Armelagos, Marshall Joseph Becker, Don Brothwell, Andrea Drusini, Marie Clabeaux Geise, Marc A. Kelley, Iwataro Moritoto, Alan G. Morris, George T. Nurse, Mary Lucas Powell, Bruce M. Rothschild, and Shelley R. Saunders. "The Origin and Antiquity of Syphilis: Paleopathological Diagnosis and Interpretation [and Comments and Reply]." *Current Anthropology* 29, no. 5 (1988): 703–737.

Begley, Christopher. "Ancient Mosquito Coast: Why Only Certain Material Culture Was Adopted from Outsiders." In *Southeastern*

Mesoamerica: Indigenous Interaction, Resilience, and Change, edited by Whitney A. Goodwin et al., 157–178. Louisville: University Press of Colorado, 2021.

———. "Elite Power Strategies and External Connections in Ancient Eastern Honduras." PhD diss., University of Chicago, 1999.

———. "I Study Collapsed Civilizations. Here Is My Advice for a Climate Change Apocalypse." *Lexington Herald-Leader*, September 23, 2019. www.kentucky.com/opinion/op-ed/article235384162.html/.

———. "Prepping." In "Post-Covid Fantasies," edited by Catherine Besteman, Heath Cabot, and Barak Kalir. *American Ethnologist*, October 19, 2020. https://americanethnologist.org/panel/pages/features/pandemic-diaries/post-covid-fantasies/prepping/edit.

Besteman, Catherine, Heath Cabot, and Barak Kalir. "Post-Covid Fantasies: An Introduction." *American Ethnologist*, July 27, 2020. https://americanethnologist.org/features/pandemic-diaries/post-covid-fantasies/post-covid-fantasies-an-introduction/.

Betancourt, P. P. "The Aegean and the Origins of the Sea Peoples." In *The Sea Peoples and Their World: A Reassessment*, edited by E. D. Oren, 297–303. Philadelphia: University of Pennsylvania Press, 2000.

Boxell, Levi, Matthew Gentzkow, and Jesse M. Shapiro. *Cross-Country Trends in Affective Polarization*. National Bureau of Economic Research Working Paper No. 26669 (2020).

Braswell, Geoffrey E. *The Maya and Teotihuacan: Reinterpreting Early Classic Interaction*. The Linda Schele Series in Maya and Pre-Columbian Studies. Austin: University of Texas Press, 2003.

Brueck, Hilary. "Doomsday Preppers Are Thinning Out Across the US, and It May Be Because President Trump Quiets Their Fears." *Business Insider*, August 26, 2019. www.businessinsider.com/what-are-doomsday-preppers-why-are-they-right-wing-2019-8/.

Carothers, Thomas, and Andrew O'Donohue, eds. *Democracies Divided: The Global Challenge of Political Polarization*. Washington, DC: Brookings Institution Press, 2019.

———. "How to Understand the Global Spread of Political Polarization." Carnegie Endowment for International Peace, October 1, 2019. https://carnegieendowment.org/2019/10/01/how-to-understand-global-spread-of-political-polarization-pub-79893/.

Chase, Diane Z., and Arlen F. Chase. "Framing the Maya Collapse: Continuity, Discontinuity, and Practice in the Classic to Postclassic

Southern Maya Lowlands." In *After Collapse: The Regeneration of Complex Societies*, edited by Glenn M. Schwartz and John J. Nichols, University of Arizona Press, 2006.

Clark, Ella E. *Indian Legends of the Pacific Northwest*. Berkeley: University of California Press, 1953.

Clifford, Patricia. "Why Did So Few Novels Tackle the 1918 Pandemic?" *Smithsonian*, November 2017. www.smithsonianmag.com/arts-culture/flu-novels-great-pandemic-180965205/.

Congressional Record, 111th Congress, First Session. "On the Endorsement of One Second After by William R. Fortschen." Extension of Remarks section, volume 155, no. 70 (May 7, 2009). www.congress.gov/congressional-record/2009/5/7/extensions-of-remarks-section/article/E1103-1/.

Crosby, Alfred W. *The Columbian Exchange: Biological and Cultural Consequences of 1492*. Westport, CT: Praeger Publishers, 2003.

Daley, Jason. "Lessons in the Decline of Democracy from the Ruined Roman Republic." *Smithsonian*, November 6, 2018. https://getpocket.com/explore/item/lessons-in-the-decline-of-democracy-from-the-ruined-roman-republic.

Denevan, William, ed. *The Native Population of the Americas in 1492*. Madison: University of Wisconsin Press, 1976.

Diamond, Jared. *Collapse: How Societies Choose to Fail or Succeed*. New York: Viking Press, 2005.

Dobyns, Henry F. "Disease Transfer at Contact." *Annual Review of Anthropology* 22 (1993): 273–291.

Dodd, Michael D., Amanda Balzer, Carly M. Jacobs, Michael W. Gruszczynski, Kevin B. Smith, and John R. Hibbing. "The Political Left Rolls with the Good and the Political Right Confronts the Bad: Connecting Physiology and Cognition to Preferences." *Philosophical Transactions of the Royal Society* B 367 (2012): 640–649.

Douglas, Peter M. J., Mark Pagani, Marcello A. Canuto, Mark Brenner, David A. Hodell, Timothy I. Eglinton, and Jason H. Curtis. "Drought, Agricultural Adaptation, and Sociopolitical Collapse in the Maya Lowlands." *Proceedings of the National Academy of Sciences of the United States of America* 112, no. 18 (2015): 5607–5612.

Dunbar-Ortiz, Roxanne. *Loaded: A Disarming History of the Second Amendment*. San Francisco: City Lights Publishers, 2018.

Ebert, Claire E., Julie Hoggarth, Jaime Awe, Brendan Culleton, and Douglas Kennett. "The Role of Diet in Resilience and Vulnerability to Climate Change Among Early Agricultural Communities in the Maya Lowlands." *Current Anthropology* 60, no. 4 (August 2019): 589–601.

Ehrlich, Paul R., and Anne H. Ehrlich, "Can a Collapse of Global Civilization Be Avoided?," *Philosophical Transactions of the Royal Society B* 280, no. 1754 (March 7, 2013).

Eisenstadt, S. N. "Beyond Collapse." In *The Collapse of Ancient States and Civilizations*, edited by N. Yoffee and G. L. Cowgill, 236–243. Tucson: University of Arizona Press, 1988.

Espinoza, Fatima. "Working at the Seams of Colonial Structures: Alternative Sociotechnical Infrastructures Revealed by Hurricane Maria." *Science, Technology, & Human Values*, forthcoming.

Ewing, Maura. "Do Right-to-Carry Gun Laws Make States Safer?" *Atlantic*, June 24, 2017.

Farley, Audrey. "The Apocalyptic Ideas Influencing Pence and Pompeo Could Also Power the Left." *Washington Post*, January 28, 2020. www.washingtonpost.com/outlook/2020/01/28/apocalyptic-ideas-influencing-pence-pompeo-could-also-power-left/.

Fea, John, Laura Gifford, R. Marie Griffith, and Lerone A. Martin. "Evangelicalism and Politics." *The American Historian, Religion and Politics* (2018). www.oah.org/tah/issues/2018/november/evangelicalism-and-politics/.

FitzGerald, Frances. *The Evangelicals: The Struggle to Shape America*. New York: Simon & Schuster, 2017.

Forstchen, William R. *One Second After*. New York: Forge Books, 2009.

Foster, Gwendolyn. *Hoarders, Doomsday Preppers, and the Culture of the Apocalypse*. New York: Palgrave Macmillan, 2014.

Gaille, Brandon. "14 Spectacular James Watt Quotes." BrandonGaille.com, November 23, 2016. https://brandongaille.com/14-spectacular-james-watt-quotes/.

Gault, Matthew. "This TV Movie About Nuclear War Depressed Ronald Reagan." *War Is Boring* (blog). Medium.com, February 19, 2015. https://medium.com/war-is-boring/this-tv-movie-about-nuclear-war-depressed-ronald-reagan-fb4c25a50044.

Gause, Emma. "A Critique: Jared Diamond's *Collapse* Put in Perspective." *Papers from the Institute of Archaeology* 24, no. 1 (2014): 1–7.

George, Andrew. *The Babylonian Gilgamesh Epic: Introduction, Critical Edition, and Cuneiform Texts*. New York: Oxford University Press, 2003.

Gibbon, Edward. *The History of the Decline and Fall of the Roman Empire*. 4 Volumes. London: W. Strahan and T. Cadell, 1776.

Green, Emma. "How Mike Pence's Marriage Became Fodder for the Culture Wars." *Atlantic*, March 30, 2017. www.theatlantic.com/politics/archive/2017/03/pence-wife-billy-graham-rule/521298/.

Gunther, Erna. *Klallam Folk Tales*, University of Washington Publications in Anthropology 1, no. 4 (1925): 113–170.

Henderson, A. Gwynn, and David Pollack. "Chapter 17: Kentucky." In *Native America: A State-by-State Historical Encyclopedia*, edited by Daniel S. Murphree. Volume 1, 393–440. Santa Barbara, CA: Greenwood Press, 2012.

Henderson, A. Gwynn. "Dispelling the Myth: Seventeenth- and Eighteenth-Century Indian Life in Kentucky." *Register of the Kentucky Historical Society* 90, no. 1 (1992): 1–25.

Hernandez, C. "Defensive Barricades of the Maya." In *Encyclopaedia of the History of Science, Technology, and Medicine in Non-Western Cultures*, edited by H. Selin. Dordrecht: Springer, 2014. https://doi.org/10.1007/978-94-007-3934-5_10086-1/.

Hoffman, Paul E. "Did Coosa Decline Between 1541 and 1560?" *Florida Anthropologist* 50, no. 1 (March 1997).

Horne, Gerald. *The Apocalypse of Settler Colonialism: The Roots of Slavery, White Supremacy, and Capitalism in Seventeenth-Century North America and the Caribbean*. New York: Monthly Review Press, 2018.

———. "The Apocalypse of Settler Colonialism." *Monthly Review* 69, no. 11 (2018): 1–22.

Hutson, Scott, Iliana Anconca Aragon, Miguel Covarrubias Reyna, Zachary Larsen, Katie Lukach, Shannon E. Plank, Richard E. Terry, and Willem Vanessendelft. "A Historical Processual Approach to Continuity and Change in Classic and Postclassic Yucatan." In *Beyond Collapse: Archaeological Perspectives on Resilience, Revitalization, and Transformation in Complex Societies*, edited by Ronald K. Faulseit, 287–309. Carbondale: Southern Illinois University Press, 2016.

Ingraham, Christian. "Spike in Violent Crime Follows Rise in Gun-Buying During Social Upheaval," *Washington Post*, July 15, 2020. www.washingtonpost.com/business/2020/07/15/gun-sales-jump-protests-coronavirus/.

Inomata, Takeshi, 猪俣健, Daniela Triadan, Jessica MacLellan, Melissa Burham, Kazuo Aoyama, 青山和夫, Juan Manuel Palomo, Hitoshi Yonenobu, 米延仁志, Flory Pinzón, Hiroo Nasu, and 那須浩郎. "High-Precision Radiocarbon Dating of Political Collapse and Dynastic Origins at the Maya Site of Ceibal, Guatemala." *Proceedings of the National Academy of Sciences of the United States of America* 114, no. 6 (2017): 293–298. https://doi.org/10.2307/26479210.

Inomata, Takeshi. "The Last Day of a Fortified Classic Maya Center: Archaeological Investigations at Aguateca, Guatemala." *Ancient Mesoamerica* 8, no. 2 (1997): 337–351.

Janusek, John J. "Collapse as Cultural Revolution: Power and Identity in the Tiwanaku to Pacajes Transition." In *The Foundations of Power in the Prehispanic Andes*, Archaeological Papers No. 14, edited by K. J. Vaughn, D. E. Ogburn, and C. A. Conlee, 175–209. Washington, DC: American Anthropological Association, 2004.

Jemisin, N. K. *The Fifth Season*. New York: Little, Brown & Company, 2015.

Jost, J. T., J. Glaser, A. W. Kruglanski, and F. J. Sulloway. "Political Conservatism as Motivated Social Cognition." *Psychological Bulletin* 129 (2003): 339–375.

Kanasawa, Satoshi. "Why Liberals Are More Intelligent Than Conservatives." *Psychology Today*, March 22, 2010. www.psychologytoday.com/us/blog/the-scientific-fundamentalist/201003/why-liberals-are-more-intelligent-conservatives/.

Kaplan, David. "The Darker Side of the 'Original Affluent Society.'" *Journal of Anthropological Research* 56, no. 3 (2000): 301–324.

Kelly, Casey Ryan. "Man-pocalypse: Doomsday Preppers and the Rituals of Apocalyptic Manhood." *Text and Performance Quarterly* 36, nos. 2–3 (2016): 95–114. https://dx.doi.org/10.1080/10462937.2016.1158415.

Kelton, Paul. "Avoiding the Smallpox Spirits: Colonial Epidemics and Southeastern Indian Survival." *Ethnohistory* 51, no. 1 (Winter 2004).

Kemp, Luke. "Are We on the Road to Civilisation Collapse?" BBC, February 18, 2019. www.bbc.com/future/article/20190218-are-we-on-the-road-to-civilisation-collapse/.

Kidder, Alfred. "Introduction." In *Prehistoric Southwesterners from Basketmaker to Pueblo*, edited by C. Amsden, xi–xiv. Los Angeles: Southwest Museum, 1949.

Klein, Ezra. "What Polarization Data from 9 Countries Reveals About the US." *Vox*, January 24, 2020. www.vox.com/2020/1/24/21076232/polarization-america-international-party-political.

Klein, Naomi. "The Rise of Disaster Capitalism." *Nation*, April 14, 2005. www.thenation.com/article/rise-disaster-capitalism/.

———. *The Shock Doctrine: The Rise of Disaster Capitalism*. New York: Metropolitan Books, 2007.

Klemko, Robert. "Behind the Armor: Men Seek 'Purpose' in Protecting Property Despite Charges of Racism." *Washington Post*, October 5, 2020. www.washingtonpost.com/national/behind-the-armor-men-seek-purpose-in-protecting-property-despite-charges-of-racism/2020/10/05/b8496fec-001e-11eb-9ceb-061d646d9c67_story.html.

Knowlton, Timothy. *Maya Creation Myths: Words and Worlds of the Chilam Balam*. Boulder: University Press of Colorado, 2010.

Koch, Alexander, Chris Brierley, Mark M. Maslin, and Simon L. Lewis. "Earth System Impacts of the European Arrival and Great Dying in the Americas After 1492." *Quaternary Science Reviews* 207 (March 1, 2019): 13–36.

Lenton, Timothy M., Johan Rockström, Owen Gaffney, Stefan Rahmstorf, Katherine Richardson, Will Steffen, and Hans Joachim Schellnhuber. "Climate Tipping Points—Too Risky to Bet Against: The Growing Threat of Abrupt and Irreversible Climate Changes Must Compel Political and Economic Action on Emissions." *Nature* 575 (2019): 592–595. https://doi.org/10.1038/d41586-019-03595-0.

"Malaria in Kentucky: Prevalence and Geographic Distribution." *Public Health Reports (1896–1970)* 32, no. 31 (1917): 1215–1221. www.jstor.org/stable/4574589.

Manahan, T. Kam, and Marcello A. Canuto. "Bracketing the Copan Dynasty: Late Preclassic and Early Postclassic Settlements at Copan, Honduras." *Latin American Antiquity* 20, no. 4, (2009): 553–580.

Mandel, Emily St. John. *Station Eleven*. New York: Knopf, 2014.

Manley, Jennifer. "Measles and Ancient Plagues: A Note on New Scientific Evidence." *Classical World* 107, no. 3 (2014): 393–397.

McAnany, Patricia Ann, and Norman Yoffee, eds. *Questioning Collapse: Human Resilience, Ecological Vulnerability, and the Aftermath of Empire*. New York: Cambridge University Press, 2010.

McAnany, Patricia, and Tomas Gallareta Negron. "Bellicose Rulers and Climatological Peril?: Retrofitting Twenty-First-Century Woes on

Eighth-Century Maya Society." In *Questioning Collapse: Human Resilience, Ecological Vulnerability, and the Aftermath of Empire*, edited by Patricia McAnany and Norman Yoffee, 142–175. New York: Cambridge University Press, 2010.

McCarthy, Cormac. *The Road*. New York: Alfred A. Knopf, 2006.

McSweeney, Kendra, and Oliver T. Coomes. "Climate-Related Disaster Opens a Window of Opportunity for Rural Poor in Northeastern Honduras." *Proceedings of the National Academy of Sciences of the United States of America* 108, no. 13 (2011): 5203–5208. https://doi.org/10.1073/pnas.1014123108.

Mencimer, Stephanie. "Evangelicals Love Donald Trump for Many Reasons, but One of Them Is Especially Terrifying: End Times." *Mother Jones*, January 23, 2020. www.motherjones.com/politics/2020/01/evangelicals-are-anticipating-the-end-of-the-world-and-trump-is-listening/.

Middleton, Guy D. "Nothing Lasts Forever: Environmental Discourses on the Collapse of Past Societies." *Journal of Archaeological Research* 20 (2012): 257–307.

———. *Understanding Collapse: Ancient History and Modern Myths*. New York: Cambridge University Press, 2017.

Milanich, Jerald T., and Charles Hudson. *Hernando de Soto and the Indians of Florida*. Gainesville: University Press of Florida, 1993.

Miller, Eric D. "Apocalypse Now? The Relevance of Religion for Beliefs About the End of the World." *Journal of Beliefs & Values* 33 (2012): 111–115.

Mills, Michael F. "Obamageddon: Fear, the Far Right, and the Rise of 'Doomsday' Prepping in Obama's America." *Journal of American Studies* 20 (2019): 1–30.

Momigliano, Arnaldo. "La Caduta Senza Rumore Di Un Impero Nel 476 D. C." *Annali Della Scuola Normale Superiore Di Pisa. Classe Di Lettere E Filosofia* 3, no. 2 (1973): 397–418.

Moran, William L. "Atrahasis: The Babylonian Story of the Flood." *Biblica* 52, no. 1 (1971): 51–61.

Nemett, Adam. *We Can Save Us All*. Los Angeles: Unnamed Press, 2018.

Niven, Larry, and Jerry Pournelle. *Lucifer's Hammer*. New York: Del Rey Books, 1977.

Nunn, Nathan, and Nancy Qian. "The Columbian Exchange: A History of Disease, Food, and Ideas." *Journal of Economic Perspectives* 24, no. 2 (2010): 163–188.

Palka, Joel W. "Ancient Maya Defensive Barricades, Warfare, and Site Abandonment." *Latin American Antiquity* 12, no. 4 (December 2001): 427–430.

Pool, Christopher A., and Michael L. Loughlin. "Tres Zapotes: The Evolution of a Resilient Polity in the Olmec Heartland of Mexico." In *Beyond Collapse: Archaeological Perspectives on Resilience, Revitalization, and Transformation in Complex Societies*, edited by Ronald K. Faulseit, 287–309. Carbondale: Southern Illinois University Press, 2016.

Powell, Colin. *Leadership Primer.* Presentation slides, 2006, available at www.hsdl.org/?view&did=467329.

Reagan, Ronald. *White House Diaries*, entry October 10, 1983. www.reaganfoundation.org/ronald-reagan/white-house-diaries/diary-entry-10101983/.

Roman, Sabin, Erika Palmer, and Markus Brede. "The Dynamics of Human–Environment Interactions in the Collapse of the Classic Maya." *Ecological Economics* 146 (2018): 312–324. www.sciencedirect.com/science/article/pii/S0921800917305578/.

Romeo, Nick. "How Archaeologists Discovered 23 Shipwrecks in 22 Days." *National Geographic*, July 11, 2016. www.nationalgeographic.com/adventure/article/greece-shipwrecks-discovery-fourni-ancient-diving-archaeology/.

Roser, Max. "Employment in Agriculture." Published online at OurWorldInData.org, 2013. Retrieved from https://ourworldindata.org/employment-in-agriculture.

Sahlins, Marshall. "Notes on the Original Affluent Society." In *Man the Hunter*, edited by R. B. Lee and I. DeVore, 85–89. New York: Aldine Publishing Company, 1968.

Schweitzer, Dahlia. *Going Viral: Zombies, Viruses, and the End of the World.* New Brunswick: Rutgers University Press, 2018.

Servigne, Pablo, Raphaël Stevens, Gauthier Chapelle, and Daniel Rodary. "Deep Adaptation Opens Up a Necessary Conversation About the Breakdown of Civilization," openDemocracy.net, August 3, 2020. www.opendemocracy.net/en/oureconomy/deep-adaptation-opens-necessary-conversation-about-breakdown-civilisation/.

Servigne, Pablo, and Raphaël Stevens. *How Everything Can Collapse: A Manual for Our Times*. Cambridge, UK: Polity, 2020.

Siegel, M., B. Solomon, A. Knopov, E. G. Rothman, S. W. Cronin, Z. Xuan, and D. Hemenway. "The Impact of State Firearm Laws on

Homicide Rates in Suburban and Rural Areas Compared to Large Cities in the United States, 1991–2016." *Journal of Rural Health* (March 2020): 255–265. https://doi.org/10.1111/jrh.12387.

Siegel, M., M. Pahn, Z. Xuan, E. Fleegler, and D. Hemenway. "The Impact of State Firearm Laws on Homicide and Suicide Deaths in the USA, 1991–2016: A Panel Study." *Journal of General Internal Medicine* 34, no. 2021–2028 (2019). https://doi.org/10.1007/s11606-019-04922-x/.

Smith, Marvin T. "Understanding the Protohistoric Period in the Southeast." *Revista de Archaeologia Americana* no. 23, Arqueologia Historica (2005): 215–229.

Snir, Ainit, Dani Nadel, Iris Groman-Yaroslavski, Yoel Melamed, Marcelo Sternberg, Ofer Bar-Yosef, and Ehud Weiss. "The Origin of Cultivation and Proto-Weeds, Long Before Neolithic Farming." *PLOS ONE* 10, no. 7 (2015): e0131422. https://doi.org/10.1371/journal.pone.0131422

Solnit, Rebecca. *A Paradise Built in Hell: The Extraordinary Communities That Arise in Disaster.* New York: Penguin Books, 2009.

Sontag, Susan. "The Imagination of Disaster." In *Against Interpretation, and Other Essays*, 147–158. New York: Farrar, Straus & Giroux, 1966.

Spinden, H. J. "The Population of Ancient America." *Geographical Review* 18, no. 4 (1928): 641–660. https://doi.org/10.2307/207952.

Stecchini, Livio C. "The Historical Problem of the Fall of Rome." *Journal of General Education* 5, no. 1 (1950): 57–61.

Stewart, Christopher. *Jungleland: A Mysterious Lost City, a WWII Spy, and a True Story of Deadly Adventure*. New York: HarperCollins, 2013.

Sugg, K. "The Walking Dead: Late Liberalism and Masculine Subjection in Apocalypse Fictions." *Journal of American Studies* 49, no. 4 (2015): 793–811. https://doi.org/10.1017/S0021875815001723.

Sullivan, Lawrence E. *Icanchu's Drum: An Orientation to Meaning in South American Religions*. New York: Macmillan, 1988.

Sutton, Matthew A. *American Apocalypse: A History of Modern Evangelicalism*. Cambridge, MA: Harvard University Press, 2014.

Swanton, John R. *The Indian Tribes of North America*. Bulletin #145. Washington, DC: Smithsonian Institution, Bureau of American Ethnology, 1952.

Synesius of Cyrene. *The Letters of Synesius of Cyrene*. Translated by Augustine Fitzgerald. Oxford, UK: Oxford University Press, 1926.

Tainter, Joseph A. *The Collapse of Complex Societies*. New Studies in Archaeology. Cambridge, UK: Cambridge University Press, 1988.

Webster, James W., George A. Brook, L. Bruce Railsback, Hai Cheng, R. Lawrence Edwards, Clark Alexander, and Philip P. Reeder. "Stalagmite Evidence from Belize Indicating Significant Droughts at the Time of Preclassic Abandonment, the Maya Hiatus, and the Classic Maya Collapse." *Palaeogeography, Palaeoclimatology, Palaeoecology* 250, nos. 1–4 (2007): 1–17.

Wessinger, Catherine. "Millennial Glossary." In *The Oxford Handbook of Millennialism*, edited by Catherine Wessinger. Oxford, UK: Oxford University Press, 2011.

Westenholz, Joan Goodnick, and Jeffrey H. Tigay. "The Evolution of the Gilgamesh Epic." *Journal of the American Oriental Society* 104, no. 2 (1984): 370–372. https://doi.org/10.2307/602207.

Wesson, Cameron B. "De Soto (Probably) Never Slept Here: Archaeology, Memory, Myth, and Social Identity." *International Journal of Historical Archaeology* 16, no. 2 (2012): 418–435.

Wilcox, Michael. *The Pueblo Revolt: An Indigenous Archaeology of Contact*. Berkeley: University of California Press, 2009.

Willis, Andy. "I, Too, Have a Christian Perspective and I Think Coal Deserves More Respect." *Lexington Herald-Leader*, August 28, 2020.

Worth, John E., Elizabeth D. Benchley, Janet R. Lloyd, and Jennifer A. Melcher. "The Discovery and Exploration of Tristán de Luna y Arellano's 1559–1561 Settlement on Pensacola Bay." *Historical Archaeology* 54, no. 2 (2020): 472–501. https://doi.org/10.1007/s41636-020-00240-w.

Wright, Lori E., and Christine D. White. "Human Biology in the Classic Maya Collapse: Evidence from Paleopathology and Paleodiet." *Journal of World Prehistory* 10, no. 2 (1996): 147–198. www.jstor.org/stable/25801093.

Zarmakoupi, Mantha. "Hellenistic & Roman Delos: The City & Its *Emporion*." *Archaeological Reports* 61 (2015): 115–132. https://doi.org/10.1017/S0570608415000125.

NOTES

INTRODUCTION

1. Christopher Begley, "I Study Collapsed Civilizations. Here Is My Advice for a Climate Change Apocalypse," *Lexington Herald-Leader*, September 23, 2019, www.kentucky.com/opinion/op-ed/article235384162.html/.

PART I—THE PAST
CHAPTER 1: WHEN THINGS FALL APART

1. Emma Gause, "A Critique: Jared Diamond's *Collapse* Put in Perspective." *Papers from the Institute of Archaeology* 24, no. 1 (2014): Art. 16, https://doi.org/10.5334/pia.467.

2. Very good examples include Guy D. Middleton, *Understanding Collapse: Ancient History and Modern Myths* (New York: Cambridge University Press, 2017), and eds. Patricia A. McAnany and Norman Yoffee, *Questioning Collapse: Human Resilience, Ecological Vulnerability, and the Aftermath of Empire* (New York: Cambridge University Press, 2010).

3. The volume of archaeological publications dedicated to the concept of collapse, or to particular events labeled "collapses" is enormous. Readers who want to explore this issue further might start with the following publications that have been among the most informative for me: Ronald K. Faulseit, ed., *Beyond Collapse: Archaeological Perspectives on Resilience, Revitalization, and Transformation in Complex Societies* (Carbondale: Southern Illinois University Press, 2016); Patricia A. McAnany and Norman Yoffee, eds., *Questioning Collapse: Human Resilience, Ecological Vulnerability, and the Aftermath of Empire* (Carbondale: Cambridge University Press, 2010); Glenn M. Schwartz, and John J. Nichols, eds.,

After Collapse: The Regeneration of Complex Societies (Tuscon: University of Arizona Press, 2006); and Joseph A. Tainter, *The Collapse of Complex Societies* (New York: Cambridge University Press, 1988).

4. Philip P. Betancourt, "The Aegean and the Origins of the Sea Peoples," in *The Sea Peoples and Their World: A Reassessment*, ed. E. D. Oren (Philadelphia: University of Pennsylvania Press, 2000); John J. Janusek, "Collapse as Cultural Revolution: Power and Identity in the Tiwanaku to Pacajes Transition," in *The Foundations of Power in the Prehispanic Andes*, Archaeological Papers No. 14, eds. K. J. Vaughn, D. E. Ogburn, and C. A. Conlee (Washington, DC: American Anthropological Association, 2004), 175–209.

5. This notion of our explanations of the past intersecting with contemporary concerns has been voiced by archaeologists for decades, for example in Richard R. Wilk, "The Ancient Maya and the Political Present," *Journal of Anthropological Research* 41 (1985): 307–326.

6. Guy D. Middleton, "Nothing Lasts Forever: Environmental Discourses on the Collapse of Past Societies," *Journal of Archaeological Research* 20 (2012): 257–307.

7. Joel W. Palka, "Ancient Maya Defensive Barricades, Warfare, and Site Abandonment," *Latin American Antiquity* 12, no. 4 (December 2001): 427–430; C. Hernandez, "Defensive Barricades of the Maya," in *Encyclopaedia of the History of Science, Technology, and Medicine in Non-Western Cultures*, ed. H. Selin (Dordrecht: Springer, 2014), https://doi.org/10.1007/978-94-007-3934-5_10086-1; Takeshi Inomata, "The Last Day of a Fortified Classic Maya Center: Archaeological Investigations at Aguateca, Guatemala," *Ancient Mesoamerica* 8, no. 2 (1997): 337–351.

8. Christopher Begley. "Ancient Mosquito Coast: Why Only Certain Material Culture Was Adopted from Outsiders," in *Southeastern Mesoamerica: Indigenous Interaction, Resilience, and Change*, ed. Whitney A. Goodwin et al. (Louisville: University Press of Colorado, 2021), 157–178.

9. Scott Hutson et al., "A Historical Processual Approach to Continuity and Change in Classic and Postclassic Yucatan," in *Beyond Collapse: Archaeological Perspectives on Resilience, Revitalization, and Transformation in Complex Societies*, ed. Ronald K. Faulseit (Carbondale: Southern Illinois University Press, 2016), 287–19.

10. H. J. Spinden, "The Population of Ancient America," *Geographical Review* 18, no. 4 (1928): 641–660, https://doi.org/10.2307/207952.

11. Jared Diamond, *Collapse: How Societies Choose to Fail or Succeed* (New York: Viking Press, 2005).

12. Some examples include Justine Shaw, "Climate Change and Deforestation: Implications for the Maya Collapse," *Ancient Mesoamerica* 14 (2003): 157–167 and Elliot M. Abrams and David J. Rue, "The Causes and Consequences of Deforestation Among the Prehistoric Maya," *Human Ecology* 16 (1988): 377–395.

13. Cameron L. McNeil, David A. Burney, and Lida Pigott Burney, "Evidence Disputing Deforestation as the Cause for the Collapse of the Ancient Maya Polity of Copan, Honduras." *Proceedings of the National Academy of Sciences of the United States of America* 107, no. 3 (2010): 1017–1022; Richard E. W. Adams, "The Collapse of Maya Civilization: A Review of Preview Theories," in *The Classic Maya Collapse*, ed. T. Patrick Culbert (Albuquerque: University of New Mexico Press, 1973), 21–34.

14. One excellent critique of several books about societal collapse is Joseph Tainter, "Collapse, Sustainability, and the Environment: How Authors Choose to Fail or Succeed," *Reviews in Anthropology* 37, no. 4 (2008): 342–371, https://doi.org/10.1080/00938150802398677/.

15. Nick Romeo, "How Archaeologists Discovered 23 Shipwrecks in 22 Days," *National Geographic*, July 11, 2016, www.nationalgeographic.com/adventure/article/greece-shipwrecks-discovery-fourni-ancient-diving-archaeology.

16. Edward Gibbon, *The History of the Decline and Fall of the Roman Empire*, 6 vols. (London: W. Strahan and T. Cadell, 1776).

17. Readers who want to explore further the decline of the Western Roman Empire might want to look at the following texts that influenced my understanding of this era, including Simon Esmonde Cleary, *The Roman West, AD 200–500: An Archaeological Study* (Cambridge, UK: Cambridge University Press, 2013); Peter Heather, *The Fall of the Roman Empire: A New History of Rome and the Barbarians* (New York: Oxford University Press, 2006); Stephen Mitchell, *A History of the Later Roman Empire, AD 284–641* (Malden, MA: John Wiley & Sons, 2014); and Chris Wickham, *Framing the Early Middle Ages: Europe and the Mediterranean, 400–800* (New York: Oxford University Press, 2005).

18. Ainit Snir et al., "The Origin of Cultivation and Proto-Weeds, Long Before Neolithic Farming," *PLOS ONE* 10 no. 7 (2015): e0131422, https://doi.org/10.1371/journal.pone.0131422.

19. John R. Swanton, *The Indian Tribes of North America*, Bulletin 145 (Washington, DC: Smithsonian Institution, Bureau of American Ethnology, 1952).

20. William Denevan, ed., *The Native Population of the Americas in 1492* (Madison: University of Wisconsin Press, 1976).

21. Chris Brierley et al., "Earth System Impacts of the European Arrival and Great Dying in the Americas After 1492," *Quaternary Science Reviews* 207 (March 2019): 13–36.

22. A. Gwynn Henderson, "Dispelling the Myth: Seventeenth- and Eighteenth-Century Indian Life in Kentucky," *Register of the Kentucky Historical Society* 90, no. 1 (1992): 1–25.

23. Michael Wilcox, *The Pueblo Revolt: An Indigenous Archaeology of Contact* (Berkeley: University of California Press, 2009).

24. Gerald Horne. "The Apocalypse of Settler Colonialism," *Monthly Review* 69, no. 11 (2018): 1–22; Gerald Horne, *The Apocalypse of Settler Colonialism: The Roots of Slavery, White Supremacy, and Capitalism in Seventeenth-Century North America and the Caribbean* (New York: Monthly Review Press, 2018).

CHAPTER 2: WHY THINGS FALL APART

1. Adam Nemett, *We Can Save Us All* (Los Angeles: Unnamed Press, 2018).

2. Adam Nemett writes in depth about our experience at the survival expo in Adam Nemett, "Journal of a Progressive Prepper: What Happens When a Homesteading Experiment Collides with a Global Pandemic?" *Rolling Stone*, June 2021, www.rollingstone.com/culture/culture-features/pandemic-prepper-homesteader-journal-1167658/.

3. James J. Aimers, "What Maya Collapse? Terminal Classic Variation in the Maya Lowlands," *Journal of Archaeological Research* 15, no. 4 (2007): 329–377.

4. Lori E. Wright and Christine D. White, "Human Biology in the Classic Maya Collapse: Evidence from Paleopathology and Paleodiet," *Journal of World Prehistory* 10, no. 2 (1996): 147–198, www.jstor.org/stable/25801093.

5. Marie Charlotte Arnauld, Chloé Andrieu, and Mélanie Forné, "'In the Days of My Life': Elite Activity and Interactions in the Maya Lowlands from Classic to Early Postclassic Times (The Long Ninth Century, AD 760–920)," *Journal de la Société des Américanistes* 103 (2017): 41–96, www.jstor.org/stable/26606790.

6. Many articles discuss the post-collapse Maya world, but I returned to these: T. Kam Manahan and Marcello A. Canuto, "Bracketing the Copan Dynasty: Late Preclassic and Early Postclassic Settlements at Copan, Honduras." *Latin American Antiquity* 20, no. 4 (2009): 553–580 and Diane Z. Chase and Arlen F. Chase, "Framing the Maya Collapse: Continuity, Discontinuity, and Practice in the Classic to Postclassic Southern Maya Lowlands," in *After Collapse: The Regeneration of Complex Societies*, ed. Glenn M. Schwartz and John J. Nichols (Tucson: University of Arizona Press, 2006).

7. Takeshi Inomata et al., "High-Precision Radiocarbon Dating of Political Collapse and Dynastic Origins at the Maya Site of Ceibal, Guatemala," *Proceedings of the National Academy of Sciences of the United States of America* 114, no. 6 (2017): 293–298, https://doi.org/10.2307/26479210.

8. Peter M. J. Douglas et al., "Drought, Agricultural Adaptation, and Sociopolitical Collapse in the Maya Lowlands," *Proceedings of the National Academy of Sciences of the United States of America* 112, no. 18 (2015): 5607–5612.

9. Claire E. Ebert et al., "The Role of Diet in Resilience and Vulnerability to Climate Change Among Early Agricultural Communities in the Maya Lowlands," *Current Anthropology* 60, no. 4 (August 2019): 589–601.

10. Geoffrey E. Braswell, *The Maya and Teotihuacan: Reinterpreting Early Classic Interaction*, Linda Schele Series in Maya and Pre-Columbian Studies (Austin: University of Texas Press, 2003).

11. James W. Webster et al., "Stalagmite Evidence from Belize Indicating Significant Droughts at the Time of Preclassic Abandonment, the Maya Hiatus, and the Classic Maya Collapse," *Palaeogeography, Palaeoclimatology, Palaeoecology* 250, nos. 1–4 (2007): 1–17.

12. Sabin Roman, Erika Palmer, and Markus Brede, "The Dynamics of Human–Environment Interactions in the Collapse of the Classic Maya," *Ecological Economics* 146 (2018): 312–324, www.sciencedirect.com/science/article/pii/S0921800917305578.

13. Gyles Iannone, ed., *The Great Maya Droughts in Cultural Context: Case Studies in Resilience and Vulnerability* (Louisville: University Press of Colorado, 2014).

14. Patricia McAnany and Tomas Gallareta Negron, "Bellicose Rulers and Climatological Peril?: Retrofitting Twenty-First-Century Woes on Eighth-Century Maya Society," in *Questioning Collapse: Human*

Resilience, Ecological Vulnerability, and the Aftermath of Empire, eds. Patricia McAnany and Norman Yoffee, 142–175 (New York: Cambridge University Press, 2010).

15. James Aimers and David Hodell, "Societal Collapse: Drought and the Maya," *Nature* 479 (2011): 44–45, https://doi.org/10.1038/479044a.

16. Good articles that summarize and critique theories of collapse include Marilyn A. Masson, "Maya Collapse Cycles," *Proceedings of the National Academy of Sciences of the United States of America* 109, no. 45 (2012): 18237–18238 and B. L. Turner and Jeremy A. Sabloff, "Classic Period Collapse of the Central Maya Lowlands: Insights about Human-Environment Relationships for Sustainability," *Proceedings of the National Academy of Sciences of the United States of America* 109, no. 35 (2012): 13908–13914.

17. The focus on systemic collapse, rather than some particular cause, has deep roots in archaeology. Publications from the 1970s came to similar conclusions, including T. Patrick Culbert, ed., *The Classic Maya Collapse*. School of American Research Books. (Albuquerque: University of New Mexico Press, 1973).

18. Mantha Zarmakoupi, "Hellenistic & Roman Delos: The City & Its *Emporion*," *Archaeological Reports* 61 (2015): 115–132, https://doi.org/10.1017/S0570608415000125.

19. Livio C. Stecchini, "The Historical Problem of the Fall of Rome," *Journal of General Education* 5, no. 1 (1950): 57–61.

20. Jennifer Manley, "Measles and Ancient Plagues: A Note on New Scientific Evidence," *The Classical World* 107, no. 3 (2014): 393–397.

21. Jason Daley, "Lessons in the Decline of Democracy from the Ruined Roman Republic," *Smithsonian*, November 6, 2018, https://getpocket.com/explore/item/lessons-in-the-decline-of-democracy-from-the-ruined-roman-republic.

22. Jerald T. Milanich and Charles Hudson, *Hernando de Soto and the Indians of Florida* (Gainesville: University Press of Florida, 1993).

23. Cameron B. Wesson, "De Soto (Probably) Never Slept Here: Archaeology, Memory, Myth, and Social Identity," *International Journal of Historical Archaeology* 16, no. 2 (2012): 418–435.

24. John E. Worth et al., "The Discovery and Exploration of Tristán de Luna y Arellano's 1559–1561 Settlement on Pensacola Bay," *Historical Archaeology* 54, no. 2 (2020): 472–501, https://doi.org/10.1007/s41636-020-00240-w.

25. Paul E. Hoffman, "Did Coosa Decline Between 1541 and 1560?" *The Florida Anthropologist* 50, no. 1 (March 1997).

26. Paul Kelton, "Avoiding the Smallpox Spirits: Colonial Epidemics and Southeastern Indian Survival," *Ethnohistory* 51, no. 1 (Winter 2004).

27. Marvin T. Smith, "Understanding the Protohistoric Period in the Southeast," *Revista de Archaeologia Americana* no. 23, Arqueologia Historica (2005): 215–229.

28. Brenda J. Baker et al., "The Origin and Antiquity of Syphilis: Paleopathological Diagnosis and Interpretation [and Comments and Reply]," *Current Anthropology* 29, no. 5 (1988): 703–737.

29. Alfred W. Crosby, *The Columbian Exchange: Biological and Cultural Consequences of 1492* (Westport, CT: Praeger Publishers, 2003); William M. Denevan, "Introduction," in *The Native Population of the Americas in 1492*, ed. William M. Denevan (Madison: University of Wisconsin Press, 1976), 1–12; Henry F. Dobyns, "Disease Transfer at Contact," *Annual Review of Anthropology* 22 (1993): 273–291; and Nathan Nunn and Nancy Qian, "The Columbian Exchange: A History of Disease, Food, and Ideas," *Journal of Economic Perspectives* 24, no. 2 (2010): 163–188.

30. A. Gwynn Henderson and David Pollack, "Chapter 17: Kentucky," in *Native America: A State-by-State Historical Encyclopedia*, vol. 1, ed. Daniel S. Murphree (Santa Barbara, CA: Greenwood Press, 2012), 393–440.

31. Henderson and Pollack, "Kentucky," 415.

CHAPTER 3: HOW THINGS FALL APART

1. "Malaria in Kentucky: Prevalence and Geographic Distribution," *Public Health Reports (1896–1970)* 32, no. 31 (1917): 1215–1221, www.jstor.org/stable/4574589.

2. An entire edited volume is dedicated to just this type of contextualization. See Geoffrey E. Braswell, ed., *The Maya and Their Central American Neighbors: Settlement Patterns, Architecture, Hieroglyphic Texts, and Ceramics* (New York: Routledge/Taylor & Francis Group, 2014).

3. Christopher Begley, "Elite Power Strategies and External Connections in Ancient Eastern Honduras" (PhD diss., University of Chicago, 1999).

4. Arnaldo Momigliano, "La Caduta Senza Rumore Di Un Impero Nel 476 D. C.," *Annali Della Scuola Normale Superiore Di Pisa. Classe Di Lettere E Filosofia* 3, no. 2 (1973): 397–418.

5. Synesius of Cyrene, *The Letters of Synesius of Cyrene*, trans. Augustine Fitzgerald (Oxford, UK: Oxford University Press, 1926).

6. Alexander Koch et al., "Earth System Impacts of the European Arrival and Great Dying in the Americas After 1492," *Quaternary Science Reviews* 207 (March 2019): 13–36.

7. Dahlia Schweitzer, *Going Viral: Zombies, Viruses, and the End of the World* (New Brunswick: Rutgers University Press, 2018).

8. S. N. Eisenstadt, "Beyond Collapse," in *The Collapse of Ancient States and Civilizations*, eds. N. Yoffee and G. L. Cowgill (Tucson: University of Arizona Press, 1988), 236–243.

9. Joseph A. Tainter, *The Collapse of Complex Societies*, New Studies in Archaeology (Cambridge, UK: Cambridge University Press, 1988).

10. Patricia Ann McAnany and Norman Yoffee, eds., *Questioning Collapse: Human Resilience, Ecological Vulnerability, and the Aftermath of Empire* (New York: Cambridge University Press, 2010).

11. Guy D. Middleton, "Nothing Lasts Forever: Environmental Discourses on the Collapse of Past Societies," *Journal of Archaeological Research* 20 (2012): 257–307.

PART II—THE PRESENT

CHAPTER 4: APOCALYPTIC FANTASIES

1. Gwendolyn Foster, *Hoarders, Doomsday Preppers, and the Culture of the Apocalypse* (New York: Palgrave Macmillan, 2014).

2. Bianca Spriggs used this term during an interview for my radio segment *Future Tense*, available at https://esweku.org/track/2367716/future-tense-with-chris-begley.

3. Patricia Clifford, "Why Did So Few Novels Tackle the 1918 Pandemic?" *Smithsonian*, November 2017, www.smithsonianmag.com/arts-culture/flu-novels-great-pandemic-180965205.

4. Roxanne Dunbar-Ortiz, *Loaded: A Disarming History of the Second Amendment* (San Francisco: City Lights Publishers, 2018), chapter 5.

5. Dunbar-Ortiz, *Loaded*, chapter 5.

6. Brian Aldiss, *Billion Year Spree: The True History of Science Fiction* (New York: Doubleday, 1973).

7. Casey Ryan Kelly, "Man-pocalypse: Doomsday Preppers and the Rituals of Apocalyptic Manhood," *Text and Performance Quarterly* 36, nos. 2–3 (2016): 95–114, http://dx.doi.org/10.1080/10462937.2016.1158415.

8. Alfred Kidder, "Introduction," in *Prehistoric Southwesterners from Basketmaker to Pueblo*, ed. C. Amsden (Los Angeles: Southwest Museum, 1949), xi–xiv.

9. This is discussed in many journalistic articles, such as: Sable Aradia, "Sexism in the Apocalypse," *Between the Shadows* blog, September 1, 2017, www.patheos.com/blogs/betweentheshadows/2017/09/sexism-in-the-apocalypse/.

10. K. Sugg, "The Walking Dead: Late Liberalism and Masculine Subjection in Apocalypse Fictions," *Journal of American Studies* 49, no. 4 (2015): 793–811, https://doi.org/10.1017/S0021875815001723.

11. Christian Ingraham, "Spike in Violent Crime Follows Rise in Gun-Buying During Social Upheaval," *Washington Post*, July 15, 2020, www.washingtonpost.com/business/2020/07/15/gun-sales-jump-protests-coronavirus/.

CHAPTER 5: APOCALYPTIC FEARS

1. Susan Sontag, "The Imagination of Disaster," in *Against Interpretation, and Other Essays* (New York: Farrar, Straus & Giroux, 1966), 157.

2. Sontag, "The Imagination of Disaster," 158.

3. Congressional Record, 111th Congress, First Session, "On the Endorsement of *One Second After* by William R. Fortschen," Extension of Remarks section, vol. 155, no. 70 (May 7, 2009), www.congress.gov/congressional-record/2009/5/7/extensions-of-remarks-section/article/E1103-1.

4. Ronald Reagan, *White House Diaries*, entry October 10, 1983, online at www.reaganfoundation.org/ronald-reagan/white-house-diaries/diary-entry-10101983/. Reagan's remarks are discussed in Matthew Gault, "This TV Movie About Nuclear War Depressed Ronald Reagan," *War Is Boring* blog, Medium.com, 2015, https://medium.com/war-is-boring/this-tv-movie-about-nuclear-war-depressed-ronald-reagan-fb4c25a50045.

5. Hilary Brueck, "Doomsday Preppers Are Thinning Out Across the US, and It May Be Because President Trump Quiets Their Fears," *Business Insider*, August 26, 2019, www.businessinsider.com/what-are-doomsday-preppers-why-are-they-right-wing-2019-8/.

6. J. T. Jost et al., "Political Conservatism as Motivated Social Cognition," *Psychological Bulletin* 129 (2003): 339–375.

7. Gwendolyn Foster, *Hoarders, Doomsday Preppers, and the Culture of the Apocalypse* (New York: Palgrave Macmillan, 2014).

8. Michael F. Mills, "Obamageddon: Fear, the Far Right, and the Rise of 'Doomsday' Prepping in Obama's America," *Journal of American Studies* 20 (2019): 1–30.

9. Levi Boxell, Matthew Gentzkow, and Jesse M. Shapiro, *Cross-Country Trends in Affective Polarization*, National Bureau of Economic Research Working Paper No. 26669, January 2020, https://doi.org/10.3386/w26669. See also Ezra Klein, "What Polarization Data from 9 Countries Reveals About the US," *Vox*, January 24, 2020, www.vox.com/2020/1/24/21076232/polarization-america-international-party-political.

10. Thomas Carothers and Andrew O'Donohue, "How to Understand the Global Spread of Political Polarization," Carnegie Endowment for International Peace, October 1, 2019, https://carnegieendowment.org/2019/10/01/how-to-understand-global-spread-of-political-polarization-pub-79893.

11. Thomas Carothers and Andrew O'Donohue, eds., *Democracies Divided: The Global Challenge of Political Polarization* (Washington, DC: Brookings Institution Press, 2019).

12. J. T. Jost et al., "Political Conservatism as Motivated Social Cognition," *Psychological Bulletin* 129 (2003): 339–375.

13. Michael D. Dodd et al., "The Political Left Rolls with the Good and the Political Right Confronts the Bad: Connecting Physiology and Cognition to Preferences," *Philosophical Transactions of the Royal Society* B 367 (2012): 640–649.

14. Satoshi Kanasawa, "Why Liberals Are More Intelligent Than Conservatives," *Psychology Today*, March 22, 2010, www.psychologytoday.com/us/blog/the-scientific-fundamentalist/201003/why-liberals-are-more-intelligent-conservatives.

15. Dahlia Schweitzer, *Going Viral: Zombies, Viruses, and the End of the World* (New Brunswick, NJ: Rutgers University Press, 2018).

16. My interview with Dahlia Schweitzer can be streamed at https://esweku.org/track/2411461/chris-begley-w-dahlia-schweitzer.

17. Rebecca Solnit, *A Paradise Built in Hell: The Extraordinary Communities That Arise in Disaster* (New York: Penguin Books, 2009).

18. M. Siegel et al., "The Impact of State Firearm Laws on Homicide Rates in Suburban and Rural Areas Compared to Large Cities in the United States, 1991–2016," *Journal of Rural Health* 36, no. 2 (March 2020): 255–265, https://doi.org/10.1111/jrh.12387; Michael Siegel et al., "The Impact of State Firearm Laws on Homicide and Suicide Deaths in the USA, 1991–2016: A Panel Study," *Journal of General Internal*

Medicine 34 (2019), https://doi.org/10.1007/s11606-019-04922-x/; Maura Ewing, "Do Right-to-Carry Gun Laws Make States Safer?" *Atlantic*, June 24, 2017.

19. Robert Klemko, "Behind the Armor: Men Seek 'Purpose' in Protecting Property Despite Charges of Racism," *Washington Post*, October 5, 2020, www.washingtonpost.com/national/behind-the-armor-men-seek-purpose-in-protecting-property-despite-charges-of-racism/2020/10/05/b8496fec-001e-11eb-9ceb-061d646d9c67_story.html/.

20. Safia Samee Ali, "Where Protesters Go, Armed Militias, Vigilantes Likely to Follow with Little to Stop Them," NBC News, September 1, 2020, www.nbcnews.com/news/us-news/where-protesters-go-armed-militias-vigilantes-likely-follow-little-stop-n1238769/.

CHAPTER 6: READING AND WRITING THE APOCALYPSE

1. Timothy Knowlton, *Maya Creation Myths: Words and Worlds of the Chilam Balam* (Boulder: University Press of Colorado, 2010).

2. For an extensive discussion and comparison of Amazonian religious narratives, see Lawrence E. Sullivan, *Icanchu's Drum: An Orientation to Meaning in South American Religions* (New York: Macmillan, 1988). For some discussion on Native American mythology, see Ella E. Clark, *Indian Legends of the Pacific Northwest* (Berkeley: University of California Press, 1953).

3. William L. Moran, "Atrahasis: The Babylonian Story of the Flood," *Biblica* 52, no. 1 (1971): 51–61.

4. Andrew George, *The Babylonian Gilgamesh Epic: Introduction, Critical Edition and Cuneiform Texts* (New York: Oxford University Press, 2003); Joan Goodnick Westenholz and Jeffrey H. Tigay, "The Evolution of the Gilgamesh Epic," *Journal of the American Oriental Society* 104, no. 2 (1984): 370–372, https://doi.org/10.2307/602207.

5. Stephanie Mencimer, "Evangelicals Love Donald Trump for Many Reasons, but One of Them Is Especially Terrifying: End Times," *Mother Jones*, January 23, 2020, www.motherjones.com/politics/2020/01/evangelicals-are-anticipating-the-end-of-the-world-and-trump-is-listening/.

6. Matthew A. Sutton, *American Apocalypse: A History of Modern Evangelicalism* (Cambridge, MA: Harvard University Press, 2014), www.jstor.org/stable/j.ctt9qdt92.

7. Frances FitzGerald, *The Evangelicals: The Struggle to Shape America* (New York: Simon & Schuster, 2017).

8. John Fea et al., "Evangelicalism and Politics," The American Historian, Religion and Politics, November 2018, www.oah.org/tah/issues/2018/november/evangelicalism-and-politics/.

9. Audrey Farley, "The Apocalyptic Ideas Influencing Pence and Pompeo Could Also Power the Left," *Washington Post*, January 28, 2020, www.washingtonpost.com/outlook/2020/01/28/apocalyptic-ideas-influencing-pence-pompeo-could-also-power-left/.

10. Drollinger's views are available through multiple publications of Capitol Ministries at capmin.org, including "Better Understanding the Believer's Future Judgment," *Members Bible Study*, March 21, 2016, http://capmin.org/wp-content/uploads/2016/03/Better-Understanding-Believers-Future-Judgment.pdf.

11. Brandon Gaille, "14 Spectacular James Watt Quotes," BrandonGaille.com (blog), November 23, 2016, https://brandongaille.com/14-spectacular-james-watt-quotes/.

12. Dale Russakoff, "James Watt and the Wages of Influence," *Washington Post*, May 4, 1989; Michael Lacey, "The Earth's Storm Troopers," *Phoenix New Times*, August 7, 1991.

13. Andy Willis, "I, Too, Have a Christian Perspective and I Think Coal Deserves More Respect," *Lexington Herald-Leader* August 28, 2020.

14. Karla Ward, "'Take a Step Back.' Beshear's Plan to Quarantine Easter Churchgoers Draws Fire from GOP," *Lexington Herald-Leader*, April 11, 2020.

15. Catherine Wessinger, "Millennial Glossary," in *The Oxford Handbook of Millennialism*, ed. Catherine Wessinger (Oxford, UK: Oxford University Press, 2011).

16. Eric D. Miller, "Apocalypse Now? The Relevance of Religion for Beliefs About the End of the World," *Journal of Beliefs & Values* 33 (2012): 111–115.

17. Emma Green, "How Mike Pence's Marriage Became Fodder for the Culture Wars," *Atlantic*, March 30, 2017, www.theatlantic.com/politics/archive/2017/03/pence-wife-billy-graham-rule/521298/.

PART III—THE FUTURE

1. Catherine Besteman, Heath Cabot, and Barak Kalir, "Post-Covid Fantasies: An Introduction," in "Post-Covid Fantasies," eds. Catherine Besteman, Heath Cabot, and Barak Kalir, *American Ethnologist*, July

27, 2020, https://americanethnologist.org/features/pandemic-diaries/post-covid-fantasies/post-covid-fantasies-an-introduction

2. Christopher Begley, "Prepping," in "Post-Covid Fantasies," eds. Catherine Besteman, Heath Cabot, and Barak Kalir, *American Ethnologist*, October 19, 2020, https://americanethnologist.org/panel/pages/features/pandemic-diaries/post-covid-fantasies/prepping/edit.

CHAPTER 7: LIKELY SCENARIOS

1. Pablo Servigne et al., "Deep Adaptation Opens Up a Necessary Conversation About the Breakdown of Civilization," OpenDemocracy, August 3, 2020, www.opendemocracy.net/en/oureconomy/deep-adaptation-opens-necessary-conversation-about-breakdown-civilisation/.

2. Paul R. Ehrlich and Anne H. Ehrlich, "Can a Collapse of Global Civilization Be Avoided?," *Philosophical Transactions of the Royal Society B* 280, no. 1754 (March 7, 2013), https://doi.org/10.1098/rspb.2012.2845; Pablo Servigne and Raphaël Stevens, *How Everything Can Collapse: A Manual for Our Times*. (Cambridge, UK: Polity, 2020).

3. Luke Kemp, "Are We on the Road to Civilisation Collapse?," BBC, February 18, 2019, www.bbc.com/future/article/20190218-are-we-on-the-road-to-civilisation-collapse/.

4. Patricia McAnany and Tomas Gallareta Negron, "Bellicose Rulers and Climatological Peril?: Retrofitting Twenty-First-Century Woes on Eighth-Century Maya Society," in *Questioning Collapse: Human Resilience, Ecological Vulnerability, and the Aftermath of Empire*, eds. Patricia McAnany and Norman Yoffee (New York: Cambridge University Press, 2010), 142–175.

5. SafeHome Team, "Best and Worst States for Climate Change," SafeHome, December 11, 2019, www.safehome.org/climate-change-statistics/; "Is Your State at Risk?" States at Risk, https://statesatrisk.org/Kentucky.

6. Timothy M. Lenton et al., "Climate Tipping Points—Too Risky to Bet Against: The Growing Threat of Abrupt and Irreversible Climate Changes Must Compel Political and Economic Action on Emissions," *Nature* 575 (2019): 592–595, https://doi.org/10.1038/d41586-019-03595-0/.

7. Naomi Klein, *The Shock Doctrine: The Rise of Disaster Capitalism* (New York: Metropolitan Books, 2007).

8. Kendra McSweeney and Oliver T. Coomes, "Climate-Related Disaster Opens a Window of Opportunity for Rural Poor in Northeastern Honduras," *Proceedings of the National Academy of Sciences of the United*

States of America 108, no. 13 (2011): 5203–5208, https://doi.org/10.1073/pnas.1014123108.

9. Naomi Klein, "The Rise of Disaster Capitalism," *Nation*, April 14, 2005, www.thenation.com/article/rise-disaster-capitalism/.

10. Fatima Espinoza, "Working at the Seams of Colonial Structures: Alternative Sociotechnical Infrastructures Revealed by Hurricane Maria," *Science, Technology & Human Values*, forthcoming.

11. McDonald Observatory, University of Texas at Austin, "What Is the Chance of Earth Being Hit by a Comet or Asteroid?" StarDate, 2021, https://stardate.org/astro-guide/faqs/what-chance-earth-being-hit-comet-or-asteroid/.

12. Christopher Begley, "I Study Collapsed Civilizations. Here Is My Advice for a Climate Change Apocalypse," *Lexington Herald-Leader*, September 23, 2019, www.kentucky.com/opinion/op-ed/article235384162.html/.

CHAPTER 8: WHO SURVIVES AND WHY

1. Chris Pool and Michael Loughlin, "Tres Zapotes: The Evolution of a Resilient Polity in the Olmec Heartland of Mexico," in *Beyond Collapse: Archaeological Perspectives on Resilience, Revitalization, and Transformation in Complex Societies*, ed. Ronald K. Faulseit (Carbondale: Southern Illinois University Press, 2016), 287–309.

2. Max Roser, "Employment in Agriculture," OurWorldInData, 2013, https://ourworldindata.org/employment-in-agriculture/.

3. Marshall Sahlins, "Notes on the Original Affluent Society," in *Man the Hunter*, eds. R. B. Lee and I. DeVore (New York: Aldine Publishing Company, 1968), 85–89.

4. David Kaplan, "The Darker Side of the 'Original Affluent Society,'" *Journal of Anthropological Research* 56, no. 3 (2000): 301–324.

5. Adam Katzenstein, "Let Me Interrupt Your Expertise with My Confidence," cartoon in *New Yorker*, January 18, 2018.

6. Jeanette Winter, *Follow the Drinking Gourd* (Decorah, IA: Dragonfly Books, 1992).

7. Christopher Stewart, *Jungleland: A Mysterious Lost City, a WWII Spy, and a True Story of Deadly Adventure* (New York: HarperCollins, 2013).

8. Colin Powell, *Leadership Primer*, PowerPoint slides, 2006, available at www.hsdl.org/?view&did=467329.

9. Fatima Espinoza, "Working at the Seams of Colonial Structures: Alternative Sociotechnical Infrastructures Revealed by Hurricane Maria," *Science, Technology & Human Values*, forthcoming.

CONCLUSION

1. Rebecca Solnit, *A Paradise Built in Hell: The Extraordinary Communities That Arise in Disaster* (New York: Penguin Books, 2009).

2. National Sustainable Agriculture Coalition, "2012 Census Drilldown: Minority and Women Farmers," NSAC's Blog, June 3, 2014, https://sustainableagriculture.net/blog/census-drilldown-sda/.

INDEX

Alicia Calton

Chris Begley is an underwater archaeologist, a wilderness survival instructor, and an anthropology professor. He completed his doctorate at the University of Chicago, was Fulbright scholar to El Salvador, and is a National Geographic Explorer. He has worked in North, Central, and South America and around the Mediterranean. He lives in Lexington, Kentucky.